LONDON MATHEMATICAL SOCIETY LECTURE NOTE SERIES

Managing Editor: Professor J.W.S. Cassels, Department of Pure Mathematics and Mathematical Statistics, University of Cambridge, 16 Mill Lane, Cambridge CB2 1SB, England

The books in the series listed below are available from booksellers, or, in case of difficulty, from Cambridge University Press.

4 Algebraic topology, J.F. ADAMS
17 Differential germs and catastrophes, Th. BROCKER & L. LANDER
27 Skew field constructions, P.M. COHN
34 Representation theory of Lie groups, M.F. ATIYAH *et al*
36 Homological group theory, C.T.C. WALL (ed)
39 Affine sets and affine groups, D.G. NORTHCOTT
40 Introduction to H_p spaces, P.J. KOOSIS
42 Topics in the theory of group presentations, D.L. JOHNSON
43 Graphs, codes and designs, P.J. CAMERON & J.H. VAN LINT
45 Recursion theory: its generalisations and applications, F.R. DRAKE & S.S. WAINER (eds)
46 p-adic analysis: a short course on recent work, N. KOBLITZ
49 Finite geometries and designs, P. CAMERON, J.W.P. HIRSCHFELD & D.R. HUGHES (eds)
50 Commutator calculus and groups of homotopy classes, H.J. BAUES
51 Synthetic differential geometry, A. KOCK
54 Markov processes and related problems of analysis, E.B. DYNKIN
57 Techniques of geometric topology, R.A. FENN
58 Singularities of smooth functions and maps, J.A. MARTINET
59 Applicable differential geometry, M. CRAMPIN & F.A.E. PIRANI
60 Integrable systems, S.P. NOVIKOV *et al*
62 Economics for mathematicians, J.W.S. CASSELS
65 Several complex variables and complex manifolds I, M.J. FIELD
66 Several complex variables and complex manifolds II, M.J. FIELD
68 Complex algebraic surfaces, A. BEAUVILLE
69 Representation theory, I.M. GELFAND *et al*
74 Symmetric designs: an algebraic approach, E.S. LANDER
76 Spectral theory of linear differential operators and comparison algebras, H.O. CORDES
77 Isolated singular points on complete intersections, E.J.N. LOOIJENGA
78 A primer on Riemann surfaces, A.F. BEARDON
79 Probability, statistics and analysis, J.F.C. KINGMAN & G.E.H. REUTER (eds)
80 Introduction to the representation theory of compact and locally compact groups, A. ROBERT
81 Skew fields, P.K. DRAXL
82 Surveys in combinatorics, E.K. LLOYD (ed)
83 Homogeneous structures on Riemannian manifolds, F. TRICERRI & L. VANHECKE
84 Finite group algebras and their modules, P. LANDROCK
85 Solitons, P.G. DRAZIN
86 Topological topics, I.M. JAMES (ed)
87 Surveys in set theory, A.R.D. MATHIAS (ed)
88 FPF ring theory, C. FAITH & S. PAGE
89 An F-space sampler, N.J. KALTON, N.T. PECK & J.W. ROBERTS
90 Polytopes and symmetry, S.A. ROBERTSON
91 Classgroups of group rings, M.J. TAYLOR
92 Representation of rings over skew fields, A.H. SCHOFIELD
93 Aspects of topology, I.M. JAMES & E.H. KRONHEIMER (eds)
94 Representations of general linear groups, G.D. JAMES
95 Low-dimensional topology 1982, R.A. FENN (ed)

96 Diophantine equations over function fields, R.C. MASON
97 Varieties of constructive mathematics, D.S. BRIDGES & F. RICHMAN
98 Localization in Noetherian rings, A.V. JATEGAONKAR
99 Methods of differential geometry in algebraic topology, M. KAROUBI & C. LERUSTE
100 Stopping time techniques for analysts and probabilists, L. EGGHE
101 Groups and geometry, ROGER C. LYNDON
103 Surveys in combinatorics 1985, I. ANDERSON (ed)
104 Elliptic structures on 3-manifolds, C.B. THOMAS
105 A local spectral theory for closed operators, I. ERDELYI & WANG SHENGWANG
106 Syzygies, E.G. EVANS & P. GRIFFITH
107 Compactification of Siegel moduli schemes, C-L. CHAI
108 Some topics in graph theory, H.P. YAP
109 Diophantine Analysis, J. LOXTON & A. VAN DER POORTEN (eds)
110 An introduction to surreal numbers, H. GONSHOR
111 Analytical and geometric aspects of hyperbolic space, D.B.A.EPSTEIN (ed)
112 Low-dimensional topology and Kleinian groups, D.B.A. EPSTEIN (ed)
113 Lectures on the asymptotic theory of ideals, D. REES
114 Lectures on Bochner-Riesz means, K.M. DAVIS & Y-C. CHANG
115 An introduction to independence for analysts, H.G. DALES & W.H. WOODIN
116 Representations of algebras, P.J. WEBB (ed)
117 Homotopy theory, E. REES & J.D.S. JONES (eds)
118 Skew linear groups, M. SHIRVANI & B. WEHRFRITZ
119 Triangulated categories in the representation theory of finite-dimensional algebras, D. HAPPEL
120 Lectures on Fermat varieties, T. SHIODA
121 Proceedings of *Groups - St Andrews 1985*, E. ROBERTSON & C. CAMPBELL (eds)
122 Non-classical continuum mechanics, R.J. KNOPS & A.A. LACEY (eds)
123 Surveys in combinatorics 1987, C. WHITEHEAD (ed)
124 Lie groupoids and Lie algebroids in differential geometry, K. MACKENZIE
125 Commutator theory for congruence modular varieties, R. FREESE & R. MCKENZIE
126 Van der Corput's method for exponential sums, S.W. GRAHAM & G. KOLESNIK
127 New directions in dynamical systems, T.J. BEDFORD & J.W. SWIFT (eds)
128 Descriptive set theory and the structure of sets of uniqueness, A.S. KECHRIS & A. LOUVEAU
129 The subgroup structure of the finite classical groups, P.B. KLEIDMAN & M.W.LIEBECK
130 Model theory and modules, M. PREST
131 Algebraic, extremal & metric combinatorics, M-M. DEZA, P. FRANKL & I.G. ROSENBERG (eds)
132 Whitehead groups of finite groups, ROBERT OLIVER
133 Linear algebraic monoids, MOHAN S. PUTCHA
134 Number thoery and dynamical systems, M. DODSON & J. VICKERS (eds)
135 Operator algebras and applications, 1, D. EVANS & M. TAKESAKI (eds)
136 Operator algebras and applications, 2, D. EVANS & M. TAKESAKI (eds)

London Mathematical Society Lecture Note Series. 135

Operator Algebras and Applications

Volume I: Structure Theory; K-Theory, Geometry and Topology

Edited by

DAVID E. EVANS
University of Wales, Swansea

MASAMICHI TAKESAKI
University of California, Los Angeles

CAMBRIDGE UNIVERSITY PRESS
Cambridge
New York New Rochelle Melbourne Sydney

Published by the Press Syndicate of the University of Cambridge
The Pitt Building, Trumpington Street, Cambridge CB2 1RP
32 East 57th Street, New York, NY 10022, USA
10, Stamford Road, Oakleigh, Melbourne 3166, Australia

First published 1988

Printed in Great Britain at the University Press, Cambridge

Library of Congress cataloging in publication data available

British Library cataloguing in publication data available

ISBN 0 521 36843 X

Preface

A symposium was organised by D.E. Evans at the Mathematics Institute, University of Warwick, between 1st October 1986 and 29th October 1987, with support from the Science and Engineering Research Council, on operator algebras and applications and connections with topology and geometry (K–theory, index theory, foliations, differentiable structures, braids, links) with mathematical physics (statistical mechanics and quantum field theory) and topological dynamics.

As part of that programme, a UK-US Joint Seminar on Operator Algebras was held during 20-25 July 1987 at Warwick, with support from SERC and NSF and organised by D.E. Evans and M. Takesaki. These two volumes contains papers, both research and expository articles, from members of that special week, together with some articles by D.B. Abraham, A.L. Carey, and A. Wassermann on work discussed earlier in the year.

We would like to take this opportunity to thank SERC and NSF for their support, and the participants, speakers and authors for their contributions.

D.E. Evans
Department of Mathematics & Computer Science
University College of Swansea
Singleton Park
Swansea SA2 8PP
Wales, U.K.

M. Takesaki
Department of Mathematics
University of California at Los Angeles
405 Hilgard Avenue
California 90024
U.S.A.

Contents

Volume I: Structure Theory; K-Theory, Geometry and Topology.

UK-US Joint Seminar on Operator Algebras
Lecture list

K-theory for discrete groups
P. Baum, A. Connes 1

Comparison theory for simple C-algebras*
B. Blackadar 21

Interpolation for multipliers
L.G. Brown 55

Elliptic invariants and operator algebras: toroidal examples
R.G. Douglas 61

On multilinear double commutant theorems
E.G. Effros, R. Exel 81

Loop spaces, cyclic homology and the Chern character
E. Getzler, J.D.S. Jones, S.B. Petrack 95

The Weyl theorem and block decompositions
R.V. Kadison 109

Secondary invariants for elliptic operators and operator algebras
J. Kaminker 119

Inverse limits of C-algebras and applications*
N.C. Phillips 127

Partitioning non-compact manifolds and the dual Toeplitz problem
J. Roe 187

Cyclic cohomology of algebras of smooth functions on orbifolds
A.J. Wassermann 229

Volume II: Mathematical Physics and Subfactors.

Some recent results for the planar Ising model
D.B. Abraham 1

The heat semigroup, derivations and Reynolds' identity
C.J.K. Batty, O. Bratteli, D.W. Robinson 23

C-algebras in solid state physics: 2D electrons in uniform magnetic field*
J. Bellissard 49

Spin groups, infinite dimensional Clifford algebras and applications
A.L. Carey 77

Subfactors and related topics
V.F.R. Jones 103

Quantized groups, string algebras, and Galois theory for algebras
A. Ocneanu 119

On amenability in type II_1 factors
S. Popa 173

*An index for semigroups of *-endomorphisms of B(H)*
R.T. Powers 185

Coactions and Yang-Baxter equations for ergodic actions and subfactors
A.J. Wassermann 203

Derived link invariants and subfactors
H. Wenzl 237

UK-US Joint Seminar on Operator Algebras

Lectures

H. Araki	*Invariant Indefinite Metric for Group Representations.*
W.B. Arveson	*Connections for Semigroups of Endomorphisms of B(H).*
C.J.K. Batty	*State Spaces, Extensions and Decomposition.*
P. Baum	*Chern Character for Discrete Groups.*
J.V. Bellissard	*Using C*–algebras in Solid State Physics.*
B. Blackadar	*Comparison Theory in Simple C*-Algebras.*
L.G. Brown	*Interpolation for Multipliers.*
R.A. Douglas	*Elliptic Invariants and Operator Algebras.*
R.L. Hudson	*Quantum Stochastic Calculus and Dilations.*
S. Hurder	*Analytic Invariants for Foliations and their Applications.*
J.D.S. Jones	*Cyclic Cohomology, Loop Spaces and the Chern Character.*
V.F.R. Jones	*Subfactors and Related Topics.*
J.T. Lewis	*Bose-Einstein Condensation and Large Deviations.*
R. Longo	*Injective Subfactors Invariant under Compact Actions.*
T.A. Loring	*Embeddings into AF Algebras and Filtrations of K-theory.*
A. Ocneanu	*On the Classification of Subfactors: Strings and Galois Theory.*
C. Phillips	*Inverse Limits of C*-Algebras and Applications.*
S. Popa	*I. On a General Johnson-Parrott Problem.*
	II. Rigidity and Amenability in Type II_1 Factors.
R.T. Powers	*An Index for Continuous Semi-Groups of *–Endomorphisms of B(H).*
J. Roe	*Cutting Manifolds in Half and Cyclic Cohomology.*
K. Schmidt	*Automorphisms of Compact Groups.*
V.G. Turaev	*Yang Baxter Equations and Link Invariants.*
A. Wassermann	*Yang-Baxter Equations for Ergodic Actions and Subfactors.*
H. Wenzl	*On the Structure of Brauer's Centralizer Algebras.*
J.D.M. Wright	*Monotone Complete C*–algebras.*

K-Theory For Discrete Groups

Paul Baum,
Department of Mathematics, Brown University,
Providence, R.I. 1917, U.S.A.
Alain Connes,
I.H.E.S., 91440 Bures-sur-Yvette, France.

Let X be a C^∞ manifold.

Suppose given

(1) A C^∞ foliation F of X

or (2) A C^∞ (right) action of a Lie group G on X,
$X \times G \to X$

or (3) A C^∞ (right) action of a discrete group
Γ on X, $X \times \Gamma \to X$.

In (2) G is a Lie group with $\pi_0(G)$ finite.
In (3) "discrete" simply means that Γ is a group topologized by the discrete topology in which each point is an open set.

Each of these three cases gives rise to a C^* algebra A

(1) $A = C^*(X, F)$, the foliation C^* algebra [10] [12]

(2) $A = C_0(X) \rtimes G$, the reduced crossed-product C^*-algebra resulting from the action of G on $C_0(X)$. As usual $C_0(X)$ denotes the abelian C^* algebra of all continuous complex-valued functions on X which vanish at infinity.

(3) $A = C_0(X) \rtimes \Gamma$, the reduced crossed-product C^*-algebra resulting from the action of Γ on $C_0(X)$.

In [4] [6] we defined for each case (and also for crossed-products twisted

---

This research was partially supported by the National Science Foundation, IHES, and the US–France Cooperative Science Program.

by a 2-cocycle) a geometric K theory and a map μ from the geometric K theory to the K theory of the relevant C^*-algebra:

(1) $\mu : K^i(X,F) \to K_i[C^*(X,F)]$

(2) $\mu : K^i(X,F) \to K_i[C_0(X) \rtimes G)]$

(3) $\mu : K^i(X,\Gamma) \to K_i[C_0(X) \rtimes \Gamma]$ $i = 0, 1$.

The map μ in essence assigns to a symbol the index of its underlying elliptic operator. In [4] [6] we conjectured that μ is always an isomorphism. At the present time this conjecture is still alive.

Over the long run it will, of course, be fascinating to see whether such an extremely general conjecture can endure.

In this note we shall briefly review the state of the art for the special case of a discrete group Γ operating on a point. The C^*-algebra is then $C^*_r\Gamma$, the reduced C^*-algebra of Γ .

On the positive side there is the beautiful Mayer-Vietoris exact sequence of M. Pimsner [27]. This exact sequence includes earlier results of Pimsner–Voiculescu [28], T. Natsume [26], C. Lance [23], J. Cuntz [14], and J. Anderson-W. Paschke [2]. Using his bivariant KK theory, G.G. Kasparov [19] [20] [21], has verified the conjecture for discrete subgroups of $SO_0(n,1)$. Also on the positive side is the transverse fundamental class result of [13]. We shall report further on this in [7].

On the negative side there are the recent examples of G. Skandalis [39]. These examples at least indicate that life is not as simple as one might have hoped. Also on the negative side is the utter lack of a Kunneth theorem and of any calculated examples where the discrete group Γ is infinite and has Kazhdan's property T. In addition there is the somewhat disturbing observation of M. Gromov that mathematicians have never proved a non-trivial theorem about all discrete groups.

It is a pleasure to thank J. Anderson, D. Burghelea, M. Gromov, J. Mingo, and G. Skandalis for enlightening and informative conversations. Also, we thank our British hosts for very lively and stimulating meetings at the Universities of Durham and Warwick during the spring and summer of 1987.

§1. Statement of the Conjecture

For simplicity assume that Γ is countable. (For Γ uncountable see Appendix 2 of [6].) $\mathfrak{C}(\cdot,\Gamma)$ denotes the category of all proper C^∞ Γ–manifolds. Thus an object of $\mathfrak{C}(\cdot,\Gamma)$ is a C^∞ manifold W with a given proper (right) C^∞ action of Γ :

(1.1) $$W \times \Gamma \to W$$

W is an ordinary C^∞ manifold. W is Hausdorff, finite dimensional, second countable, and without boundary. Each $\gamma \in \Gamma$ acts on W by a diffeomorphism. Recall that the action (1.1) is proper if and only if the map

(1.2) $$W \times \Gamma \to W \times W$$

which takes (w,γ) to $(w,w\gamma)$ is proper in the usual sense, i.e. the inverse image of any compact set in $W \times W$ is a compact set in $W \times \Gamma$. This implies that each isotropy group of the action (1.1) is finite, and that the quotient space W/Γ is a Hausdorff second countable orbifold.

A morphism in $\mathfrak{C}(\cdot,\Gamma)$ is a C^∞ Γ-equivariant map $f : W_1 \to W_2$. f is not required to be proper. Associated to f is a homomorphism of abelian groups

(1.3) $$f_! : K_i[C_0(TW_1) \rtimes \Gamma] \to K_i[C_0(TW_2) \rtimes \Gamma]$$
$$i = 0,\, 1$$

In (1.3) TW_j is the tangent bundle of W_j . $C_0(TW_j) \rtimes \Gamma$ is the crossed-product C^*-algebra resulting from the evident action of Γ on TW_j . To define $f_!$ first note that TW_j is an even-dimensional almost-complex manifold. The derivative map

(1.4) $$f' : TW_1 \to TW_2$$

is, therefore, a K-oriented map and so yields

$$(f') \in KK^0(C_0(TW_1)\,,\, C_0(TW_2))\,.$$

Consider the diagram

$$KK^0_\Gamma(C_0(TW_1), C_0(TW_2)) \to KK^0(C_0(TW_1), C_0 TW_2))$$

$$\downarrow$$

$$KK^0(C_0(TW_1) \rtimes \Gamma, C_0(TW_2) \rtimes \Gamma)$$

$$\downarrow$$

$$\mathrm{Hom}_{\mathbb{Z}}(K_*[C_0(TW_1) \rtimes \Gamma], K_*[C_0(TW_2) \rtimes \Gamma])$$

in which the horizontal arrow is the forgetful map and the two vertical arrows are as in Kasparov [19] [20]. Since all structures involved in defining

$(f') \in KK^0(C_0(TW_1), C_0(TW_2))$ are Γ-equivariant (f') lifts to give

$[f'] \in KK^0_\Gamma(C_0(TW_1), C_0(TW_2))$. The horizontal arrow in (1.5) sends $[f']$ to (f'). The map $f_!$ of (1.3) is then obtained by applying the two vertical arrows to $[f']$.

(1.6) ***Lemma:*** $f_! : K_i[C_0(TW_1) \rtimes \Gamma] \to K_i[C_0(T(W_2) \rtimes \Gamma]$ depends only on the homotopy class (as a C^∞ Γ-equivariant map) of f.

(1.7) ***Lemma:*** Let $f : W_1 \to W_2$ and $g : W_2 \to W_3$ be morphisms in $\mathcal{C}(\cdot,\Gamma)$. Then $(gf)_! = g_! f_!$.

(1.8) ***Definiton:*** $$K^i(\cdot,\Gamma) = \underset{\mathcal{C}(\cdot,\Gamma)}{\xrightarrow{\text{limit}}} K_i[C_0(TW) \rtimes \Gamma] \quad i = 0,1$$

Remarks: In (1.8) the limit is taken using the $f_!$ maps of (1.3). Let $F_i(\cdot,\Gamma)$ be the free abelian groups generated by all[1] pairs (W, ξ) where W is an object of $\mathcal{C}(\cdot,\Gamma)$ and $\xi \in K_i[C_0(TW) \rtimes \Gamma]$. $R_i(\cdot,\Gamma)$ denotes the subgroups of $F_i(\cdot,\Gamma)$ generated by all elements of the form

(i) $(W, \xi+\eta) - (W,\xi) - (W,\eta)$

(ii) $(W_1,\xi) - (W_2, f_!\xi)$

Then definition (1.8) is:

(1.9) $$K^i(\cdot,\Gamma) = F_i(\cdot,\Gamma) / R_i(\cdot,\Gamma) \quad i = 0, 1$$

For each object W of Γ we have the homomorphism of abelian groups

(1.10) $$\varphi_W : K_i[C_0(TW) \rtimes \Gamma] \to K^i(\cdot,\Gamma)$$

defined by

(1.11) $$\varphi_W(\xi) = (W, \xi) \qquad \xi \in K_i[C_0(TW) \rtimes \Gamma]$$

For any morphism $f : W_1 \to W_2$ in $\mathcal{C}(\cdot,\Gamma)$, the diagram

[1] To avoid set-theoretic difficulties assume that W is a C^∞ sub-manifold (which is a closed subset) of some Euclidean space $\mathbb{R}^m$. This is possible by the Whitney embedding theorem.

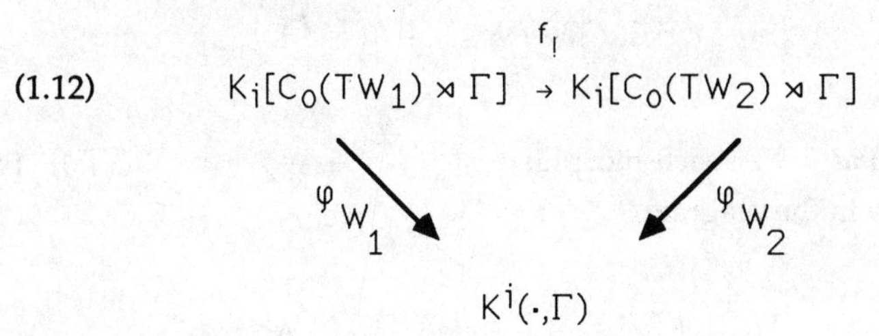

commutes.

$K^i(\cdot,\Gamma)$ has the following universal property. Let H be an abelian group. Suppose given for each object W of $\mathfrak{C}(\cdot,\Gamma)$ a homomorphism of abelian groups

(1.13) $$\psi_W : K_i[C_0(TW) \rtimes \Gamma] \to H$$

Assume that for each morphism $f : W_1 \to W_2$ in $\mathfrak{C}(\cdot,\Gamma)$ the diagram

(1.14) $$\begin{array}{ccc} K_i[C_0(TW_1) \rtimes \Gamma] & \xrightarrow{f_!} & K_i[C_0(TW_2) \rtimes \Gamma] \\ & {\scriptstyle \psi_{W_1}} \searrow \quad \swarrow {\scriptstyle \psi_{W_2}} & \\ & H & \end{array}$$

commutes.

Then there is a unique homomorphism of abelian groups

(1.15) $$\psi : K^i(\cdot,\Gamma) \to H$$

with $\psi_W = \psi\, \varphi_W$ for each object W of $\mathfrak{C}(\cdot,\Gamma)$.

(1.15) applies to determine a homomorphism

(1.16) $$\mu : K^i(\cdot,\Gamma) \to K_i\,[C^*_r\Gamma]\,.$$

For each object W of $\mathfrak{C}(\cdot,\Gamma)$ the Dirac operator of TW is an element of $KK^0_\Gamma(C_0(TW),\mathbb{C})$. Applying Kasparov's map

(1.17) $$KK^0_\Gamma(C_0(TW),\mathbb{C}) \to KK^0(C_0(TW) \rtimes \Gamma,\; C^0_r\Gamma)$$

we then obtain a homomorphism of abelian groups

(1.18) $$\mu_W : K^i[C_0(TW) \rtimes \Gamma] \to K_i[C^*_r\Gamma]$$

(1.19) ***Lemma:*** For each morphism $f : W_1 \to W_2$ in $\mathcal{C}(\cdot,\Gamma)$ there is commutativity in the diagram

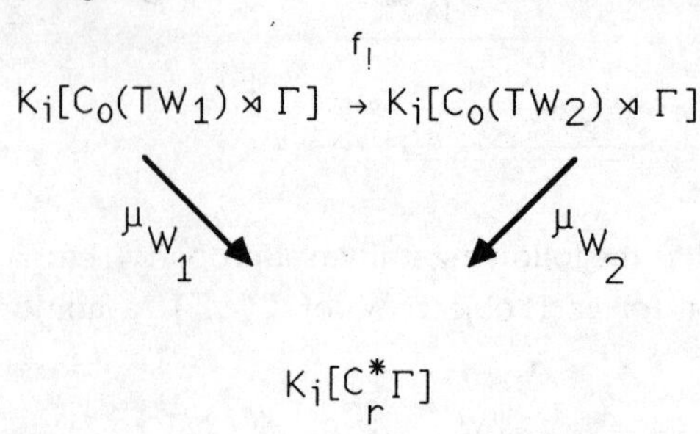

Proof: $[D_j] \in KK_\Gamma(C_0(TW_j), \mathbb{C})$ denotes the element of $KK_\Gamma(C_0(TW_j), \mathbb{C})$ given by the Dirac operator of TW_j. As above we have

$[f] \in KK^0_\Gamma(C_0(TW_1), C_0(TW_2))$. The Kasparov product pairing

$$KK^0_\Gamma(C_0(TW_1), C_0(TW_2)) \otimes_{\mathbb{Z}} KK^0_\Gamma(C_0(TW_2), \mathbb{C}) \to KK_\Gamma(C_0(TW_1), \mathbb{C})$$

has $$[f] \bigotimes_{C_0(TW_2)} [D_2] = [D_1]$$ □

Due to (1.19), (1.15) applies to determine a homomorphism

$$\mu : K^i(\cdot,\Gamma) \to K_i[C^*_r\Gamma]$$

(1.20) ***Conjecture*** $\mathbb{Z}$. For any (discrete) group

Γ $\mu : K^i(\cdot,\Gamma) \to K_i[C^*_r\Gamma]$ is an isomorphism of abelian groups $(i = 0,1)$.

Remarks: From a slightly heuristic point of view, conjecture (1.20) can be viewed as asserting that any element of $K_0[C^*_r\Gamma]$ is the index of a Γ–equivariant elliptic operator on a proper C^∞ Γ-manifold. The only relations imposed on these indices to obtain $K_0[C^*_r\Gamma]$, according to the conjecture, are the "obvious" index-preserving relations on the symbols.

Conjecture (1.20) appears to be quite a strong statement. It's truth implies validity of:

(i) Novikov conjecture on homotopy invariance of higher signatures [9].

(ii) Gromov - Lawson - Rosenberg conjecture on topological obstructions to the existence of Riemannian metrics of positive scalar curvature [15] [16] [33].

(iii) Kadison - Kaplansky conjecture that for Γ torsion-free there are no projections in $C^*_r\Gamma$ other then 0 and 1.

If Γ is torsion-free, then $K^i[\cdot,\Gamma] = K_i(B\Gamma)$, the i-th K-homology (with compact supports) of the classifying space $B\Gamma$.

§2. Chern Character

Let $S(\Gamma)$ be the set of all elements in Γ of finite order. The identity element of Γ is in $S(\Gamma)$.

(2.1) $$S(\Gamma) = \{\gamma \in \Gamma \mid \gamma \text{ is of finite order }\}$$

Γ acts on $S(\Gamma)$ by conjugation. $F\Gamma$ denotes the permutation module (with coefficients $\mathbb{C}$) so obtained. Thus a typical element of $F\Gamma$ is a finite formal sum

$$\sum_{\gamma \in S(\Gamma)} \lambda_\gamma[\gamma] \qquad \lambda_\gamma \in \mathbb{C}$$

In the standard way $F\Gamma$ is a vector space over $\mathbb{C}$. The (right) action of Γ on $F\Gamma$ is

$$\left(\sum_{\gamma \in S(\Gamma)} \lambda_\gamma[\lambda]\right)\alpha = \sum_{\gamma \in S(\Gamma)} \lambda_\gamma[\alpha^{-1}\gamma\alpha]$$

$$\alpha \in \Gamma$$

$$\lambda_\gamma \in \mathbb{C}$$

$H_j(\Gamma, F\Gamma)$ denotes the j-th homology group of Γ with coefficients $F\Gamma$. Let

Γ act on $\mathbb{C}$ trivially. Then

(2.2) $$H_j(\Gamma, F\Gamma) = \mathrm{Tor}^j_{\mathbb{C}(\Gamma)}(F\Gamma), \mathbb{C}) \qquad j = 0, 1, 2, \ldots .$$

In (2.3) $\mathbb{C}(\Gamma)$ is the group algebra of all finite formal sums

$$\sum_{\gamma \in S(\Gamma)} \lambda_\gamma [\gamma].$$

Set
$$H_{ev}(\Gamma, F\Gamma) = \bigoplus_j H_{2j}(\Gamma, F\Gamma)$$

$$H_{odd}(\Gamma, F\Gamma) = \bigoplus_j H_{2j+1}(\Gamma, F\Gamma)$$

There is then [6] a Chern character

$$ch : K^0(\cdot, \Gamma) \to H_{ev}(\Gamma, F\Gamma)$$
$$ch : K^1(\cdot, \Gamma) \to H_{odd}(\Gamma, F\Gamma)$$

constructed by using cyclic cohomology. Its crucial property is that it becomes an isomorphism after tensoring $K^i(\cdot, \Gamma)$ with $\mathbb{C}$. Hence

$$ch : K^0(\cdot, \Gamma) \otimes_{\mathbb{Z}} \mathbb{C} \to H_{ev}(\Gamma, F\Gamma)$$

$$ch : K^1(\cdot, \Gamma) \otimes_{\mathbb{Z}} \mathbb{C} \to H_{odd}(\Gamma, F\Gamma)$$

are isomorphisms of vector spaces over $\mathbb{C}$.

Consider the diagram

$$\begin{array}{ccc} K^i(X, \Gamma) & \xrightarrow{\mu} & K_0[C^*_r\Gamma] \\ \downarrow ch & & \downarrow \\ H^{ev}(\Gamma, F\Gamma) & \longrightarrow & K_0[C^*_r\Gamma] \otimes_{\mathbb{Z}} \mathbb{C} \end{array}$$

in which the right vertical arrow is the tautological map. The lower horizontal arrow is defined by requiring commutativety in the diagram, and is denoted

$$\mu : H_{ev}(\Gamma, F\Gamma) \to K_0[C_r^*\Gamma] \otimes_{\mathbb{Z}} \mathbb{C}$$

The same procedure determines

$$\mu : H_{odd}(\Gamma, F\Gamma) \to K_1[C_r^*\Gamma] \otimes_{\mathbb{Z}} \mathbb{C}$$

(2.4) ***Conjecture*** $\mathbb{C}$. For any group Γ

$$\mu : H_{ev}(\Gamma, F\Gamma) \to K_0[C_r^*\Gamma] \otimes_{\mathbb{Z}} \mathbb{C}$$

and

$$\mu : H_{odd}(\Gamma, F\Gamma) \to K_1[C_r^*\Gamma] \otimes_{\mathbb{Z}} \mathbb{C}$$

are isomorphisms of vector spaces over $\mathbb{C}$.

Remarks: If (1.20) is valid for a group Γ then so is (2.4). The point of (2.4) is that it relates homological invariants of Γ to the K theory of $C_r^*\Gamma$.

Let $L = \{\gamma_1, \gamma_2, \gamma_3, \ldots\}$ be a subset of $S(\Gamma)$ such that any element in $S(\Gamma)$ is conjugate to one and only one of the γ_i. $Z(\gamma_i)$ denotes the centralizer of γ_i in Γ. Let $H_j(Z(\gamma_i), \mathbb{C})$ be the j-th homology of $Z(\gamma_i)$ with coefficients $\mathbb{C}$ and trivial action of $Z(\gamma_i)$ on $\mathbb{C}$. Then for each $j = 0, 1, 2, \ldots$

$$H_j(\Gamma, F\Gamma) = \bigotimes_i H_j(Z(\gamma_i), \mathbb{C}) \tag{2.5}$$

(2.5) is useful in calculating $H_*(\Gamma, F\Gamma)$ in examples.

§3. Finite Groups and Abelian Groups

As noted by J. Rosenberg, operator algebraists tend to think that any statement involving only finite or abelian groups is trivial. This is surely debatable, but let us check (1.20) and (2.4) for such groups.

(3.1) ***Lemma:*** Conjecture $\mathbb{C}$ is valid for any finite group.

Proof: Let Γ be a finite group.

$K_1[C^*_r\Gamma] = 0$ and $H_j(\Gamma, F\Gamma) = 0$ for $j > 0$, so (2.4) is valid in the odd case. $K_0[C^*_r\Gamma] = R(\Gamma)$, the representation ring of Γ. Let $C\ell(\Gamma)$ be the $\mathbb{C}$-vector-space of all complex-valued functions on Γ which are constant on each conjugacy class.

(3.2) $\quad C\ell(\Gamma) = \{f : \Gamma \to \mathbb{C} \mid f(\gamma) = f(\alpha^{-1}\gamma\alpha) \text{ for all } \gamma, \alpha \in \Gamma\}$

Let $\chi : R(\Gamma) \to C\ell(\Gamma)$ be the map which assigns to a representation its character.

$H^0(\Gamma, F\Gamma) = C\ell(\Gamma)$ and $\chi : R(\Gamma) \otimes_{\mathbb{Z}} \mathbb{C} \to C\ell(\Gamma)$ is the inverse to

$\underline{\mu} : C\ell(\Gamma) \to R(\Gamma) \otimes_{\mathbb{Z}} \mathbb{C}$. □

(3.3) ***Lemma.*** Conjecture $\mathbb{Z}$ is valid for any finite group.

Proof: Let Γ be a finite group. Any (continuous) action of a finite group is proper. Let Γ act on a point. This is a final object in $\mathcal{C}(\cdot, \Gamma)$ and it is immediate that $K^i(\cdot, \Gamma) = K_i[C^*_r\Gamma]$. □

(3.4) ***Lemma.*** Conjecture $\mathbb{C}$ is valid for any abelian group.

Proof: Let Γ be an abelian group. $\hat{\Gamma}$ denotes the Pontryagin dual of Γ. Forget the group structure on $\hat{\Gamma}$, and view $\hat{\Gamma}$ as a compact Hausdorff topological space. Then $C^*_r\Gamma = C(\hat{\Gamma})$ and $K_i[C^*_r\Gamma] = K^i(\hat{\Gamma})$ the K-theory of $\hat{\Gamma}$ as defined topologically by Atiyah and Hirzebruch [1]. Moreover $H_j(\Gamma, F\Gamma) = H^j(\hat{\Gamma}; \mathbb{C})$ where $H^j(\hat{\Gamma}; \mathbb{C})$ is the j-th Čech cohomology group of $\hat{\Gamma}$ with coefficients the complex numbers $\mathbb{C}$ The topological Chern character

$$\mathrm{ch} : K^0(\hat{\Gamma}) \to \bigoplus_{j \in \mathbb{N}} H^{2j}(\hat{\Gamma};\mathbb{C})$$

$$\mathrm{ch} : K^1(\hat{\Gamma}) \to \bigoplus_{j \in \mathbb{N}} H^{2j+1}(\hat{\Gamma};\mathbb{C})$$

provides the inverse map to

$$\mu : H_{ev}(\Gamma, F\Gamma) \to K_0[C_r^*\Gamma] \otimes_{\mathbb{Z}} \mathbb{C}$$

$$\mu : H_{odd}(\Gamma, F\Gamma) \to K_0[C_r^*\Gamma] \otimes_{\mathbb{Z}} \mathbb{C}$$ □

(3.5) ***Lemma.*** Conjecture $\mathbb{Z}$ is valid for any abelian group.

Proof: First one checks that conjecture $\mathbb{Z}$ is valid for any finitely generated abelian group. Next, any abelian group is the direct limit of its finitely generated subgroups. Since both $K_*[C_r^*\Gamma]$ and $K^*(\cdot,\Gamma)$ commute with direct limits this completes the proof. □

§4. Evidence For The Conjecture

(4.1) ***Theorem.*** Let G be a connected simply-connected solvable Lie group. Then conjectures $\mathbb{Z}$ and $\mathbb{C}$ are valid for any discrete subgroup Γ of G.

Proof: The theorem is a corollary of the Thom isomorphism [11] for crossed-products by $\mathbb{R}$. See also [32]. □

Notation. $SO(n,1)$ denotes the Lorentz group. $SO(n,1) \subset SL(n+1,\mathbb{R})$ is the subgroup of $SL(n+1,\mathbb{R})$ which preserves the form $x_1^2 + \ldots + x_n^2 - x_{n+1}^2$.

(4.2) ***Theorem:*** (G.G. Kasparov [21]) Let G be a connected Lie group such that G is locally isomorphic to

$$H \times SO(n_1,1) \times SO(n_2,1) \times \ldots \times SO(n_\ell,1)$$

where H is a compact Lie group and $n_1, n_2, \ldots, n_\ell$ is any ℓ-tuple of positive integers. Then conjectures $\mathbb{Z}$ and $\mathbb{C}$ are valid for any discrete subgroup Γ of G.

Proof: See [20] [21]. □

The next theorem is a consequence of the Mayer-Victoris exact sequence of M. Pimsner [27].

(4.3) ***Theorem:*** Let Γ act on a tree without inversion. Assume that conjecture $\mathbb{Z}$ and $\mathbb{C}$ are valid for the stabilizer group of each vertex and each edge. Then conjectures $\mathbb{Z}$ and $\mathbb{C}$ are valid for Γ.

Proof: The six-term sequence constructed by M. Pimsner also exists for $K^*(\cdot,\Gamma)$ and $H_*(\Gamma, F\Gamma)$. The theorem then follows by the five lemma. □

M. Pimsner's remarkable result [27] contains previous results of Pimsner-Voiculescu [28], J. Cuntz [14], J. Anderson- W. Paschke [2], C. Lance [23], and T. Natsume [26].

In particular, three corollaries of (4.3) are:

(4.4) ***Corollary:*** Let Γ be a subgroup of Γ_1 and Γ_2. Set $\tilde{\Gamma} = \Gamma_1 *_{\Gamma} \Gamma_2$, the free product of Γ_1 and Γ_2 with amalgamation along the common subgroup Γ. Assume the conjectures $\mathbb{Z}$ and $\mathbb{C}$ are valid for Γ, Γ_1, and Γ_2. Then conjectures $\mathbb{Z}$ and $\mathbb{C}$ are valid for $\tilde{\Gamma}$.

(4.5) ***Corollary.*** Let Γ be a subgroup of Γ_1 and let $\theta: \Gamma \to \Gamma_1$ be an injective homomorphism of Γ into Γ_1. Let Γ be the HNN extension determined by Γ, Γ_1, θ. Assume that conjectures $\mathbb{Z}$ and $\mathbb{C}$ are valid for Γ and Γ_1. Then conjectures $\mathbb{Z}$ and $\mathbb{C}$ are valid for Γ.

(4.6) ***Corollary.*** (Pimsner-Voiculescu [28]) Let $\theta_1, \theta_2, \ldots, \theta_n$ be automorphisms of Γ. Denote the free group on n generators by F_n and let F_n act on Γ via $\theta_1, \theta_2, \ldots, \theta_n$. Form the semi-direct product $\Gamma \rtimes F_n$. Assume that conjectures $\mathbb{Z}$ and $\mathbb{C}$ are valid for Γ. Then conjectures $\mathbb{Z}$ and $\mathbb{C}$ are valid for $\Gamma \rtimes F_n$.

Remark: According to [38], Corollaries (4.4) and (4.5) together are equivalent to (4.3).

Another point in favor of the conjecture is the direct limit property:

(4.7) ***Proposition:*** Let $\{\Gamma_i\}_{i\in I}$ be subgroups of Γ such that with respect to inclusions $\Gamma_i \subset \Gamma_j$, $\Gamma = \xrightarrow{\text{limit}} \Gamma_i$. Assume that conjectures $\mathbb{Z}$ and $\mathbb{C}$ are valid for each Γ_i. Then conjectures $\mathbb{Z}$ and C are valid for Γ.

Proof: As observed in the proof of lemma (3.5)

$$K_*[C_r^*\Gamma] = \xrightarrow{\text{limit}} K_*[C_r^*\Gamma_i]$$

and

$$K^*(\cdot,\Gamma) = \xrightarrow{\text{limit}} K^*(\cdot,\Gamma_i) \qquad \square$$

For the injectivity of $\mu : K^*(\cdot,\Gamma) \to K_*[C_r^*\Gamma]$ Kasparov [20] has proven that μ is injective whenever Γ is a discrete subgroup of a connected Lie group. If Γ is any discrete group then the transverse fundamental class result of [13] yields a quite general partial injectivity result for

$$\mu : H_*(B\Gamma;\mathbb{C}) \to K_*[C_r^*\Gamma] \otimes_{\mathbb{Z}} \mathbb{C}$$

This will be reported on in [7].

§5. Negative Indications

Let G be a Lie group with $\pi_0 G$ finite. Let Γ be a discrete torsion-free subgroup of G. Denote the maximal compact subgroup of G by K and consider the C^∞ manifold $K\backslash G/\Gamma$. Let $T\Gamma$ be the tangent bundle of this manifold

(5.1) $$T\Gamma = T(K\backslash G/\Gamma)$$

Then

(5.2) $$K^i(\cdot,\Gamma) = K^i(T\Gamma)$$

In (5.2) $K^i(T\Gamma)$ is the usual Atiyah-Hirzebruch K-theory (with compact supports) of $T\Gamma$. Equivalently,

(5.3) $$K^i(\cdot,\Gamma) = K_i[C_0(T\Gamma)]$$

In this context, consider the case when $G = Sp(n,1)$, $n \geq 2$, and $\Gamma \subset Sp(n,1)$ is torsion-free and of finite co-volume. G. Skandalis [39] proves that for such a Γ, $C^*_r\Gamma$ is not K-nuclear. In particular $C^*_r\Gamma$ cannot be KK-equivalent to an abelian C^*-algebra so

$$\mu : K_i[C_0(T\Gamma)] \to K_i[C^*_r\Gamma]$$

cannot be a KK-equivalence of $C^*_r\Gamma$ and $C_0(T\Gamma)$.

This result of G. Skandalis came as something of a shock. The "dual Dirac" element of Kasparov-Miscenko [20] [25] in $KK(C^*_r\Gamma, C_0(T\Gamma))$ gives a map

(5.4) $$\beta : K_i[C^*_r\Gamma] \to K_i[C_0(T\Gamma)]$$

It was commonly believed [8] that at the KK level μ and β were inverses of each other. According to G. Skandalis this is not the case. But as homomorphisms of K-theory, μ and β may still be inverses of each other. Clearly something delicate and intereting is happening in these Skandalis examples which involves the difference between

$$K_*[C^*_{max}\Gamma] \text{ and } K_*[C^*_r\Gamma].$$

For a general discrete group Γ consider the standard map $C^*_{max}\Gamma \to C^*_r\Gamma$. Γ is said to be K-amenable if this map is a KK-equivalence. Any infinite group Γ having Kazhdan's [22] property T is not K-amenable. A severe weakness in the current evidence for conjectures $\mathbb{Z}$ and $\mathbb{C}$ is that we have no example of an infinite group Γ having property T for which the conjecture has been verified. Perhaps the first example that comes to mind is $SL(3,\mathbb{Z})$.

J. Anderson has observed that $SL(3,\mathbb{Z})$ is a property T group and that conjecture $\mathbb{C}$ asserts

$$K_0[C_r^*SL(3,\mathbb{Z})] \otimes_{\mathbb{Z}} \mathbb{C} = \mathbb{C}^8$$

$$K_1[C_r^*SL(3,\mathbb{Z})] \otimes_{\mathbb{Z}} \mathbb{C} = 0$$

Another weakness in the evidence for conjectures $\mathbb{Z}$ and $\mathbb{C}$ is the lack of a Künneth theorem for $K_*[C_r^*(\Gamma_1 \times \Gamma_2)]$. Here $\Gamma_1 \times \Gamma_2$ is the Cartesian product of Γ_1 and Γ_2

(5.5) $$\Gamma_1 \times \Gamma_2 = \{(\gamma_1, \gamma_2) | \gamma_i \in \Gamma_i\}$$

Set $K_i^{\mathbb{C}}(\Gamma) = K_i[C_r^*\Gamma] \otimes_{\mathbb{Z}} \mathbb{C}$

Then conjecture $\mathbb{C}$ implies

(5.6) $$K_0^{\mathbb{C}}(\Gamma_1 \times \Gamma_2) = K_0^{\mathbb{C}}(\Gamma_1) \otimes_{\mathbb{C}} K_0^{\mathbb{C}}(\Gamma_2) \oplus K_1^{\mathbb{C}}(\Gamma_1) \otimes_{\mathbb{C}} K_1^{\mathbb{C}}(\Gamma_2)$$

(5.7) $$K_1^{\mathbb{C}}(\Gamma_1 \times \Gamma_2) = K_1^{\mathbb{C}}(\Gamma_1) \otimes_{\mathbb{C}} K_0^{\mathbb{C}}(\Gamma_2) \oplus K_0^{\mathbb{C}}(\Gamma_1) \otimes_{\mathbb{C}} K_1^{\mathbb{C}}(\Gamma_2)$$

At the present time there is neither a proof nor a counter-example for (5.6) and (5.7).

Using topological methods S. Cappell [9] has proved that if the Novikov conjecture on homotopy invariance of higher signatures is valid for Γ_1 and Γ_2, then it is valid for $\Gamma_1 \times \Gamma_2$. This provides some indication that (5.6) and (5.7) may be true. If $C_r^*\Gamma_1$ is KK-equivalent to an abelian C^*-algebra (e.g. Γ_1 finite, abelian, or a discrete subgroup of $SO_0(n,1)$) then (5.6) and (5.7) are implied by the Künneth theorem of J. Rosenberg and C. Schochet [34] [35].

Let N be a normal subgroup of Γ so that the sequence

$$1 \longrightarrow N \longrightarrow \Gamma \longrightarrow \Gamma/N \longrightarrow 1$$

is exact. If conjectures $\mathbb{Z}$ and $\mathbb{C}$ are valid for N and Γ/N are they valid for Γ ? Our understanding of this question appears to be in an extremely primitive state. For example, let B_n be the braid group on n strings [3]. As usual B_n maps to the symmetric group S_n . The kernel is the pure braid group B^0_n

$$1 \longrightarrow B^0_n \longrightarrow B_n \longrightarrow S_n \longrightarrow 1$$

V. Jones has pointed out that by using the Pimsner-Voiculescu exact sequence [28] and an inductive argument one can verify conjectures $\mathbb{Z}$ and $\mathbb{C}$ for B^0_n. By (3.1) and (3.3) conjectures $\mathbb{Z}$ and $\mathbb{C}$ are valid for the symmetric group S_n . We do not know if conjectures $\mathbb{Z}$ and $\mathbb{C}$ are valid for B_n . According to conjecture $\mathbb{C}$

$$K_0[C^*_r B_n] \otimes_{\mathbb{Z}} \mathbb{C} = \mathbb{C}$$

$$K_1[C^*_r B_n] \otimes_{\mathbb{Z}} \mathbb{C} = \mathbb{C}$$

At the opposite extreme from discrete subgroups of Lie groups are the groups constructed by D. Kan and W. Thurston in [18]. The Kan-Thurston construction begins by selecting a "large" acyclic group Γ . For example, Γ could be the acyclic group of J. Mather [24]. This Γ is the group of all compactly supported homeomorphisms of the real line $\mathbb{R}$ onto itself. According to conjecture $\mathbb{Z}$

$$K_0[C^*_r \Gamma] = \mathbb{Z}$$

$$K_1[C^*_r \Gamma] = 0$$

Mather's acyclic Γ is torsion free. Other "large" acyclic discrete groups are studied in [17] [37] [40].

§6. The More General Conjecture

This note has been devoted to $K_*[C^*_r\Gamma]$. As pointed out in the introduction conjectures $\mathbb{Z}$ and $\mathbb{C}$ are special cases of a very general conjecture involving foliations, Lie group actions, and actions of discrete groups. For a precise formulation of the more general conjecture see [4] [6]. It is somewhat artificial to focus attention exclusively on conjectures $\mathbb{Z}$ and $\mathbb{C}$ because there is a great deal of inter-action among the various cases of the general conjecture. For example, let Γ be a discrete group acting by a C^∞ (right) action on a C^∞ manifold X. In [6] we construct a commutative diagram:

$$\begin{array}{ccc} K^i(X,\Gamma) & \xrightarrow{\mu} & K_i[C_0(X) \rtimes \Gamma] \\ \Big\downarrow ch_\Gamma & & \Big\downarrow \\ H^i(X,\Gamma) & \xrightarrow{\underline{\mu}} & K_i[C_0(X) \rtimes \Gamma] \otimes_{\mathbb{Z}} \mathbb{C} \end{array}$$

in which $C_0(X) \rtimes \Gamma$ is the reduced crossed-product C^*-algebra and the right vertical map is the tautological map.

(6.1) ***Conjecture:*** $\quad \mu : K^i(X,\Gamma) \to K_i[C_0(X) \rtimes \Gamma]$

and

$$\underline{\mu} : H^i(X,\Gamma) \to K_i[C_0(X) \rtimes \Gamma] \otimes_{\mathbb{Z}} \mathbb{C}$$

are isomorphisms.

Conjecture (6.1) is hereditary in Γ, i.e. if conjecture (6.1) is valid for any C^∞ action of a discrete group Γ, then conjecture (6.1) is valid for any C^∞ action of any subgroup H of Γ. The reason is that there is a "Shapiro's lemma" [29] [30] [31] [36]

$$K^i(X,H) \cong K^i(X \times_H \Gamma, \Gamma)$$

$$K_i[C_0(X) \rtimes H] \cong K_i[C_0(X \times_H \Gamma) \rtimes \Gamma]$$

So far any discrete group Γ for which conjectures $\mathbb{Z}$ and $\mathbb{C}$ are known to be valid is also known to satisfy conjecture (6.1) for any C^∞ action of Γ.

§7. Gromov's Principle

Gromov's Principle is: "There is no statement about all discrete groups which is both non-trivial and true." If Gromov is right, then the universe of all discrete groups is too wild and too crazy for us to say anything of substance about all discrete groups. In this case the correct strategy for conjectures $\mathbb{Z}$ and $\mathbb{C}$ is to try to prove them only for certain kinds of groups, e.g. discrete subgroups of Lie groups. But is Gromov right? Time will tell.

References

[1] M.F. Atiyah, *K-Theory* . W.A. Benjamin, New York, 1967.

[2] J. Anderson and W. Paschke, "The K-Theory of the reduced C*-algebra of an HNN-group", *J. Operator Theory* 16 (1986), 165-187.

[3] E. Artin, "Theorie der Zopfe", *Abh. Math. Sem. Hamburg Univ.* 4(1925), 42-72; "Theory of Braids", *Ann. of Math.* (2) 48 (1947),101-126.

[4] P. Baum and A. Connes, "Geometric K Theory For Lie Groups and Foliations", Brown University - *IHES preprint* , 1982.

[5] P. Baum and A. Connes, "Leafwise homotopy equivalence and rational Pontryagin classes", **Foliations**, (ed. I. Tamura) *Advanced Studies in Pure Mathematics 5* , North-Holland, Amsterdam-New York-Oxford, 1985, 1-14.

[6] P. Baum and A. Connes, "Chern Character for Discrete Groups", Penn State-IHES preprint 1986, to appear in *A Fête of Topology,* North-Holland, Amsterdam-New York-Oxford, 1987.

[7] P. Baum and A. Connes, "De Rham Cohomology for Discrete Groups", to appear.

[8] B. Blackadar, *K Theory for Operator Algebras,* MSRI publications 5, Springer-Verlag, New York, 1986.

[9] S.E. Cappell, "On homotopy invariance of higher signatures," *Invent. Math.* 33 (1976), 171-179.

[10] A. Connes, "Sur la théorie non commutative de l'intégration", *Springer lecture notes in math.* 725, Springer-Verlag, 1979,19-143.

[11] A. Connes, "An Analogue of the Thom Isomorphism for Crossed Products of a C*-Algebra by an Action of $\mathbb{R}$", Advances in Math. 39 (1981), 31-55.

[12] A. Connes, "A Survey of Foliations and Operator Algebras", **Operator algebras and applications** (ed. R.V. Kadison), *Proc. Symp. Pure Math.* 38(1982) *Amer. Math. Soc.*, Providence, Part I, 521-628.

[13] A. Connes, "Cyclic Cohomology and the Transverse FundamentalClass of a Foliation", **Geometric methods in operator algebras,** (eds. H. Araki and E.G. Effros) *Pitman research notes in mathematics series* 123 (1986), 257-360.

[14] J. Cuntz, "K-Theoretic Amenability for Discrete Groups", *J. Reine Angew. Math.* 344 (1983), 180-195.

[15] M. Gromov and H.B. Lawson, Jr., "Spin and Scalar Curvature in the Presence of a Fundamental Group, I", *Ann. of Math.* (2) 111 (1980), 209-230.

[16] M. Gromov and H.B. Lawson, Jr. "Positive Scalar Curvature and the Dirac Operator on Complete Riemannian Manifolds", *Publ Math. IHES* 58 (1983), 83-196.

[17] P. de la Harpe and D. McDuff, "Acyclic Groups of Automorphisms", *Université de Geneve preprint.*

[18] D.M. Kan and W.P. Thurston, "Every Connected Space has the Homology of a $K(\pi,1)$, *Topology* 15 (1976), 253-258.

[19] G.G. Kasparov, "The Operator K-functor and Extensions of C^*-algebras", *Izv. Akad. Nauk. SSSR, Ser. Mat.* 44 (1980), 571-636 = *Math. USSR Izv*. 16 (1981), 513-572.

[20] G.G. Kasparov, "K theory, Group C^*-algebras, and Higher Signatures (Conspectus) (2 parts), to appear.

[21] G.G. Kasparov, "Lorentz Groups: K-theory of Unitary Representations and Cross Products", *Dokl. Akad. Nauk SSSR* 275 (1984), 541-545.

[22] D. Kazhdan, "Connection of the Dual Space of a Group with the Structure of its Closed Subgroups", *Funct. Anal. Appl.* 1 (1967), 63-65.

[23] C. Lance, "K-theory for Certain Group C^*-algebras", *Acta. Math.* 151 (1983), 209-230.

[24] J. Mather, "The Vanishing of the Homology of Certain Groups of Homeomorphisms", Topology 10 (1971), 297-298.

[25] A.S. Miscenko, "Infinite Dimensional Representations of Discrete Groups and Higher Signatures" *Math. USSR Izv.* 8 (1974), 85-112.

[26] T. Natsume, "On $K_*(C^*(SL_2(\mathbb{Z})))$", *J. Operator Theory* 13 (1985), 103-118.

[27] M. Pimsner, "KK Groups of Crossed Products by Groups Acting on Trees", *Invent. Math*. 86 (1986), 603-634.

[28] M. Pimsner and D. Voiculescu, "K-Groups of Reduced Crossed Products by Free Groups", *J. Operator Theory* 8 (1982), 131-156.

[29] M.A. Rieffel, "Strong Morita Equivalence of Certain Transformation Group C^*-algebras", *Math. Ann*. 222 (1976), 7-22.

[30] M.A. Rieffel, "Morita Equivalence for Operator Algebras", **Operator Algebras and Applications** (ed. R.V. Kadison) *Proc. Symp. Pure Math.* 38 (1982) *Amer. Math. Soc.,* Providence, Part I, 285-298.

[31] M.A. Rieffel, "Applications of Strong Morita Equivalence to Transformation Group C*-algebras", **Operator Algebras and Applications** (ed. R.V. Kadison) *Proc. Symp. Pure Math.* 38 (1982) *Amer. Math. Soc., Providence,* Part I, 299-310.

[32] J. Rosenberg, "Group C^*-algebras and Topological Invariants", **Proc. Conf. On Operator Algebras and Group Representations,** *Neptun, Romania* , 1980. *Pitman, London* , 1981.

[33] J. Rosenberg, "C^*-algebras, Positive Scalar Curvature, and the Novikov Conjecture," *Publ. Math. IHES* 58 (1983), 197-212.

[34] J. Rosenberg and C. Schochet, "The Künneth Theorem and the Universal Coefficient Theorem for Kasparov's Generalized K-Functor", *MSRI pre-print.*

[35] J. Rosenberg and C. Schochet, "The Künneth Theorem and the Universal Coefficient Theorem for Equivariant K Theory and KK Theory", *Mem. Amer. Math. Soc.* Number 348, Volume 62 (1986), Amer. Math. Soc., Providence.

[36] D. Schwalbe, Ph.D. Thesis, Brown University, 1986.

[37] G. Segal, "Classifying Spaces Related to Foliations", Topology 17 (1978), 367-382.

[38] J.P. Serre, *"Trees"* , Springer-Verlag, New York, 1980.

[39] G. Skandalis, "Une notion de nucléarité en K-théorie", *preprint Université Pierre et Marie Curie, Paris.*

[40] J.B. Wagoner, "Delooping Classfying Spaces in Algebraic K-theory", *Topology* 11 (1972), 349-370.

COMPARISON THEORY FOR SIMPLE C*-ALGEBRAS

Bruce Blackadar
University of Nevada, Reno

1. INTRODUCTION

1.1. In my opinion, the most important general structure question concerning simple C*-algebras is the extent to which the Murray-von Neumann comparison theory for factors is valid in arbitrary simple C*-algebras. In this article, I will discuss this question, which I call the *Fundamental Comparability Question*, describe a reasonable framework which has been developed to study the problems involved in the question, give a summary of the rather modest progress made to date including some interesting classes of examples, and give some philosophical arguments why the problems are reasonable, interesting, and worth studying.

To simplify the discussion, in this article we will only consider *unital* C*-algebras. Most of the theory works equally well in the nonunital case, but there are some annoying technicalities which we wish to avoid here.

Much of what we do in this article will be an elaboration of certain aspects of the theory developed in [**Bl 4**,§5-6], to which the reader may refer for more information.

Before stating the problem in detail, let us introduce some basic notation and terminology:

Definition 1.1.1. Let p and q be projections in a C*-algebra A. Then p is *equivalent* to q, written $p \sim q$, if there is a partial isometry $u \in A$ with $u^*u=p$, $uu^*=q$.

p is *subordinate* to q, written $p \precsim q$, if p is equivalent to a subprojection of q.

p is *strictly subordinate* to q, written $p \prec q$, if p is equivalent to a proper subprojection of q.

Note that it is possible to have $p \prec p$: for example A=**B**(H), p=1. If $p \nprec p$, then p is *finite*; if $p \prec p$, then p is *infinite*. A is *finite* [resp. *infinite*] if 1_A is finite [resp. infinite]; A is *stably finite* if $M_n(A)$ (the $n{\times}n$-matrix algebra over A) is finite for all n. A is *purely infinite* if every nonzero projection in A is infinite.

If A is finite and p, q are projections in A, then $p \prec q$ if and only if $p \precsim q$ and $p \nsim q$.

1.2. The well-known basic result in the Murray-von Neumann theory of factors is the following:

Theorem 1.2.1. Let A be a factor, τ the standard trace on A. If p and q are projections in A with $\tau(p) \le \tau(q)$, then $p \precsim q$.

If we simply substitute the word "simple C*-algebra" for "factor", the statement of the theorem does not make sense, since a general simple C*-algebra does not have a standard trace -

it may have more than one tracial state [Bl 4,7.6.1] or none at all. However, the statement may be sensibly rephrased as follows:

1.2.2. Fundamental Comparability Question, Version 1 (FCQ1). Let A be a simple C*-algebra. If p, q are nonzero projections in A with $\tau(p) \leq \tau(q)$ for all tracial states τ on A, is $p \precsim q$?

This statement is true for every factor which is a simple C*-algebra, i.e. for finite factors and countably decomposable type III factors (in the latter case there are no tracial states, so the hypothesis is vacuous, and all nonzero projections are equivalent.)

1.3. There are, however, well-known examples of simple C*-algebras which do not satisfy the FCQ1. There are many examples of a simple AF algebra A with unique trace, containing two inequivalent projections of the same trace [Bl 4,7.6.2].

Since the difficulty appears to be with equivalence rather than strict comparability, one might hope that the following variation may be a more reasonable question:

1.3.1. Fundamental Comparability Question, Version 2 (FCQ2). Let A be a simple C*-algebra. If p, q are nonzero projections in A with $\tau(p) < \tau(q)$ for every trace (tracial state) τ on A, is $p \prec q$?

If A has no traces, then the FCQ2 just says that $p \prec q$ for any projections p, q with $q \neq 0$. If A has a trace, then the word "nonzero" may be deleted from the question.

The importance of this question, in addition to its intrinsic interest, is that in most cases it is quite easy to determine all of the tracial states on a simple C*-algebra and to calculate the values on given projections. In fact, it seems that most simple C*-algebras arising "naturally" have at most one tracial state, which is given by a simple formula. However, it is generally very hard to explicitly decide when two given projections are equivalent or comparable.

There is no known example of a simple C*-algebra not satisfying FCQ2, although the only classes of examples for which the statement is known to be true (factors, AF algebras, irrational rotation algebras, Cuntz algebras, and the like) are fairly special.

The presence of many projections makes the FCQ much easier to handle, and there are better results known for simple C*-algebras rich in projections (complete results in the case of AW*-factors.) Section 4 is concerned with existence of projections and connections with the FCQ. In simple C*-algebras without many projections, it is more appropriate to consider comparison theory and a version of the FCQ for general positive elements; such a theory is described in section 6.

1.4. There are five sub-problems contained in the FCQ. The first two are famous open questions which are really only tangentially related to the FCQ, and the other three are more central. Each will be discussed in more detail below.

1.4.1. Is every finite simple C*-algebra stably finite?

1.4.2. Does every (stably) finite simple C*-algebra have a tracial state?

1.4.3. Does every (stably) finite simple C*-algebra have strict cancellation?

1.4.4. Is every (stably) finite simple C*-algebra weakly unperforated?

1.4.5. Does every (stably) finite simple C*-algebra have enough (quasi)traces?

1.5. There are two opposing philosophies one may take with regard to the FCQ. Naively, the question seems reasonable; however, there are counterexamples to some of the sub-problems among non-simple C*-algebras, even among type I C*-algebras. Since simple infinite-dimensional C*-algebras have been traditionally regarded as "more complicated" than type I C*-algebras, the prevailing philosophy seems to have been that any pathology which occurs in the type I case must *a fortiori* occur for simple C*-algebras, so the lack of a "smoking gun" counterexample to the FCQ merely reflects our still rudimentary supply of well-understood examples of simple C*-algebras.

On the other hand, there is some philosophical evidence and a growing body of empirical evidence which suggests that the structure of simple C*-algebras may indeed be "simpler" in certain important respects than the structure of general C*-algebras or even of commutative C*-algebras. The pathologies in the problems of 1.4 seem to be topological in nature and tend to disappear as the algebra becomes "more noncommutative." And it seems reasonable to view simple C*-algebras, the most noncommutative C*-algebras of all, as being quite homogeneous, the idea being that an element can be moved around arbitrarily by unitary conjugation so that the only characteristic distinguishing one projection (or positive element) from another is its "size" rather than its "location"; it is reasonable to expect that "size" can be entirely measured by numerical invariants such as traces. So while it is still too early to confidently make conjectures about the FCQ, I believe the arguments and evidence in its favor must be taken seriously.

2. EXISTENCE OF TRACES

In this section, we examine the FCQ in the case where the algebra has no tracial states.

2.1. If A is simple and has a tracial state, then A must be stably finite. If A is not stably finite, then there are three possibilities: (1) A is finite but not stably finite; (2) A is infinite but not purely infinite; (3) A is purely infinite.

The purely infinite case is fairly easy [Cu 3,1.5]:

Theorem 2.1.1. Let A be a simple C*-algebra, p, q projections in A, with q infinite. Then $p \prec q$. So if A is purely infinite, then A satisfies the FCQ.

On the other hand, if A is infinite but not purely infinite, then A cannot satisfy the FCQ, since an infinite projection cannot be subordinate to a finite projection. So for infinite algebras, the FCQ is equivalent to the following question:

Question 2.1.2. Is every infinite simple C*-algebra purely infinite?

Question 2.1.2 is equivalent to question 1.4.1: if A is finite but not stably finite, then $M_n(A)$ is infinite but not purely infinite for some n; conversely, if A is infinite but not purely infinite, and p is a nonzero finite projection in A, then pAp is finite but not stably finite.

2.1.3. There exists a finite C*-algebra which is not stably finite [**Bl** 4,6.10.1]; this algebra (the Toeplitz algebra on the unit ball of $\mathbf{C}^2$) is even type I, an extension of the compact operators $\mathbf{K}$ by $C(S^3)$. There is no reason to expect that any similar construction will yield a simple example, since a counterexample to 1.4.1 must have a very different nature than this example. This Toeplitz algebra has the property that if p is a projection in a matrix algebra which is equivalent to 1, then $1-p$ is "small" (contained in a proper ideal.) In fact, an infinite simple C*-algebra must be "very infinite" in the sense that it must contain an infinite sequence of projections which are mutually orthogonal and all equivalent to 1 (i.e. must contain a copy of the Cuntz algebra O_∞) [**Cu 1**]. The construction discussed below in 6.6 is also relevant to this problem, and provides further philosophical evidence that a finite simple C*-algebra may always be stably finite.

2.2. Now let us turn our attention to stably finite simple C*-algebras. Question 1.4.2 is one of the oldest and most famous open problems in the theory of operator algebras (cf. [**MvN**,Chapter IV]). There has been much work on this question, which we will briefly summarize here, but not much real progress. The work done so far centers around the notion of a quasitrace:

Definition 2.2.1. A *quasitrace* on a C*-algebra A is a function $\tau: A \to \mathbf{C}$ satisfying

(1) $0 \le \tau(x^*x) = \tau(xx^*)$ for all $x \in A$

(2) $\tau(\lambda x) = \lambda\tau(x)$ for all $x \in A$, $\lambda \in \mathbf{C}$

(3) τ is linear on commutative C*-subalgebras of A

(4) τ extends to a map from $M_n(A)$ to $\mathbf{C}$ satisfying (1), (2), (3).

τ is *normalized* if $\tau(1) = 1$. Denote the set of normalized quasitraces on A by $QT(A)$, and the set of normalized linear quasitraces (tracial states) on A by $T(A)$.

$QT(A)$ is obviously a compact convex set in the topology of pointwise convergence, and $T(A)$ is a closed convex subset. It is well known that $T(A)$ is always a Choquet simplex, which is metrizable if A is separable; the same is true of $QT(A)$ by [**BH**].

The fundamental open question is:

Question 2.2.2. Is every quasitrace a trace (i.e. linear)?

Much of the importance of quasitraces lies in the following result due to Handelman [**Hd**], based on work of Cuntz [**Cu 2**]:

Theorem 2.2.3. If A is a stably finite simple C*-algebra, then $QT(A)$ is not empty.

The proof of this theorem uses the relationship between quasitraces and dimension functions, which yields additional reasons why $QT(A)$ is important in studying the FCQ. This theory will be discussed in more detail in section 6.

In fact, for any quasitrace τ on a (simple) C*-algebra A there is a GNS-type construction (using ultraproducts) whereby A can be embedded into a finite AW*-algebra M_τ, which is a factor if τ is extremal, such that τ extends to M_τ [**BH**]. Since any type II_1 AW*-factor has a unique quasitrace (this was in effect proved by Murray and von Neumann in [**MvN**] in their first attempt to prove that every finite W*-factor has a trace), we see that Questions 1.4.2 and 2.2.2 are equivalent. Also, since every type II_1 AW*-factor whose quasitrace is linear is a W*-factor [**W**], both questions are equivalent to:

Question 2.2.4. Is every finite AW*-factor a W*-factor?

A wild (non-W*) finite AW*-factor would be a counterexample to FCQ2.

There is a relevant result of Cuntz and Pedersen [**CP**]:

Theorem 2.2.5. Let A be a simple C*-algebra. Then $T(A)$ is nonempty if and only if $\sum x_i x_i^* \le \sum x_i^* x_i$ implies $\sum x_i x_i^* = \sum x_i^* x_i$. The sums may be either finite sums or norm-convergent infinite sums.

So if A is simple and has no tracial state, then there are elements $x_1, \cdots, x_n \in A$ with $\sum_{i=1}^{n} x_i x_i^* \lneqq \sum_{i=1}^{n} x_i^* x_i$.

Definition 2.2.6. Let A be a simple C*-algebra.

$$\mathrm{HN}(A) = \min \{n : \text{there exist } x_1, \cdots, x_n \in A \text{ with } \sum_{i=1}^{n} x_i x_i^* \lneqq \sum_{i=1}^{n} x_i^* x_i\}$$

$$\mathrm{HN}_1(A) = \min \{n : \text{there exist } x_1, \cdots, x_n \in A \text{ with } \sum_{i=1}^{n} x_i x_i^* \lneqq \sum_{i=1}^{n} x_i^* x_i = 1\}$$

$\mathrm{HN}_1(A) = 1$ if and only if A is infinite; $\mathrm{HN}(A) = \mathrm{HN}_1(A) = \infty$ if and only if A has a trace. $\mathrm{HN}(A) = 1$ if and only if A contains a nonnormal hyponormal element (an x with $xx^* \lneqq x^*x$.)

Proposition 2.2.7. For any A, $\mathrm{HN}(A) \le \mathrm{HN}_1(A) \le \mathrm{HN}(A) + 1$; if A is a finite AW*-factor, then $\mathrm{HN}_1(A) = \mathrm{HN}(A) + 1$.

Proof: It is trivial that $HN(A) \le HN_1(A)$. If $\sum_{i=1}^{n} x_i x_i^* \lneqq \sum_{i=1}^{n} x_i^* x_i = a$, by scaling we may assume $a \le 1$. Set $x_{n+1} = (1-a)^{1/2}$. Then $\sum_{i=1}^{n+1} x_i x_i^* \lneqq \sum_{i=1}^{n+1} x_i^* x_i = 1$. If A is a finite AW*-factor and $\sum_{i=1}^{n} x_i x_i^* \lneqq \sum_{i=1}^{n} x_i^* x_i = 1$, write $x_n = au$ with u unitary, $a \ge 0$, and let $y_i = x_i u^*$ for $1 \le i \le n$. Then

$$\sum_{i=1}^{n-1} y_i y_i^* + a^2 = \sum_{i=1}^{n} x_i x_i^* \lneqq \sum_{i=1}^{n-1} y_i^* y_i + a^2 = u(\sum_{i=1}^{n} x_i^* x_i)u^* = 1$$

so $\sum_{i=1}^{n-1} y_i y_i^* \lneqq \sum_{i=1}^{n-1} y_i^* y_i$. //

I believe the following conjecture can be proved independent of 2.2.4, and may be useful in attacking 2.2.4.

Conjecture 2.2.8. If A is a wild finite AW*-factor, then $HN(A) = 2$, i.e. A contains no nonnormal hyponormal elements but A contains two elements x, y with $xx^* + yy^* \lneqq x^*x + y^*y$.

3. COMPARABILITY IN STABLY FINITE ALGEBRAS

3.1. To avoid the pathologies possible from negative answers to 1.4.1 or 1.4.2, we rephrase the FCQ as follows:

3.1.1. Fundamental Comparability Question, Version 3 (FCQ3). Let A be a stably finite simple C*-algebra, p, q projections in A. If $\tau(p) < \tau(q)$ for all (normalized) quasitraces τ, is $p \precsim q$?

If A is stably finite and satisfies the conclusion of the FCQ3, we say that A has *strict comparability*. A has *stable strict comparability* if $M_n(A)$ has strict comparability for all n.

Question 3.1.2. Does strict comparability imply stable strict comparability?

In the future we will be exclusively concerned with stable strict comparability, and more generally in properties preserved in matrix algebras. It will be convenient to identify all matrix algebras over A as subalgebras of a common "infinite matrix algebra" over A:

Definition 3.1.3. $M_\infty(A)$ is the algebraic direct limit of the algebras $M_n(A)$ under the (nonunital) embeddings of $M_n(A)$ into $M_{n+k}(A)$ obtained by sending a to $diag\,(a, 0) = \begin{bmatrix} a & 0 \\ 0 & 0 \end{bmatrix}$.

$M_\infty(A)$ may be regarded as the algebra of all infinite matrices over A with only finitely many nonzero entries. We will freely regard $M_n(A)$ as the C*-subalgebra of $M_{n+k}(A)$ or $M_\infty(A)$ consisting of all matrices which are zero outside the $n \times n$-block in the upper left-hand corner.

Note that contrary to our standing assumption, $M_\infty(A)$ is nonunital. It is also not a C*-algebra since it is not complete (its completion is the stable algebra $A \otimes \mathbf{K}$ of A.)

If x and y are elements of matrix algebras over A, we denote by $x \oplus y$ the matrix $diag(x,y)$. There is a slight ambiguity here due to the identifications of elements in different matrix algebras, but $x \oplus y$ is well defined up to unitary equivalence (by a permutation matrix.) If p and q are orthogonal projections in $M_n(A)$, then $p + q \sim p \oplus q$.

3.2. Cancellation.

Definition 3.2.1. A C*-algebra A has *cancellation* if whenever p, q, r are projections in matrix algebras over A with $p \oplus r \sim q \oplus r$, then $p \sim q$. A has *strict cancellation* if whenever $p \oplus r \prec q \oplus r$, then $p \prec q$.

It is clear that stable strict comparability implies strict cancellation.

Proposition 3.2.2. Cancellation implies strict cancellation.

Proof: Suppose $p \oplus r \prec q \oplus r$. Then there is a projection s with $p \oplus r \oplus s \sim q \oplus r$. By cancellation $p \oplus s \sim q$, so $p \prec q$. //

Question 3.2.3. Does strict cancellation imply cancellation?

It is worth noting that cancellation can be rephrased in the following way [**Bl** 4,6.4.1]:

Proposition 3.2.4. Let A be a C*-algebra. Then A has cancellation if and only if, for every n, equivalent projections in $M_n(A)$ are unitarily equivalent in $M_n(A)$.

A C*-algebra with strict cancellation must be stably finite. A stably finite C*-algebra, even a commutative C*-algebra, need not have strict cancellation:

Example 3.2.5. Let $\mathbf{T}^n$ be the n-torus. Then if $n \geq 5$ $C(\mathbf{T}^n)$ does not have strict cancellation [**Rf** 3,8.4].

There are, however, some substantial classes of simple C*-algebras for which cancellation can be proved. We will discuss these examples later.

3.3. Perforation.

If p is a projection in a matrix algebra over A, by slight abuse of notation we write $n \cdot p = p \oplus \cdots \oplus p$ (n times.)

Definition 3.3.1. A C*-algebra A has *n-power cancellation* if whenever p, q are projections in a matrix algebra over A with $n \cdot p \sim n \cdot q$, then $p \sim q$.

A is *n-unperforated* if $n \cdot p \precsim n \cdot q$ implies $p \precsim q$.

A is *strictly n-unperforated* if $n \cdot p \prec n \cdot q$ implies $p \prec q$.

A is *weakly n-unperforated* if $n \cdot p \oplus r \prec n \cdot q \oplus r$ for some r implies $p \oplus s \prec q \oplus s$ for some s.

A has *power cancellation* [resp. is *unperforated*, etc.] if it has n-power cancellation [resp. is n-unperforated, etc.] for all n.

It is clear that A is n-unperforated if and only if A is strictly n-unperforated and has n-power cancellation.

Strict n-unperforation does not imply n-unperforation because $K_0(A)$ may have torsion even if A is simple [**Bl 4**,10.11.2]. Strict n-unperforation (for any n) implies stable finiteness; for stably finite simple C*-algebras, stable strict comparability implies strict unperforation.

Weak n-unperforation seems to be a rather strange condition when written as in 3.3.1, but it has a more natural formulation in terms of the ordered K_0-group as described below. In addition, it seems to be the aspect of perforation which is relevant to the FCQ and which is independent of cancellation questions:

Theorem 3.3.2. [**Bl 5**] Let A be a stably finite simple C*-algebra. Then, for any n, A is strictly n-unperforated if and only if A is weakly n-unperforated and has strict cancellation. (The "if" part is trivial.)

3.3.2 actually applies more generally to any C*-algebra A for which every projection in $M_\infty(A)$ is full. Such a C*-algebra is called *K-simple.* Any A for which Prim(A) contains no non-trivial compact open subsets is K-simple. In particular, $C(X)$ is K-simple if X is connected.

Theorem 3.3.2 is a corollary of the following relationship between perforation and cancellation:

Theorem 3.3.3. [**Bl 5**] Let A be a K-simple C*-algebra, p, q nonzero projections in $M_\infty(A)$. If $p \oplus r \sim q \oplus r$ [resp. $p \oplus r \prec q \oplus r$] for some r, then for all sufficiently large k, $k \cdot p \sim k \cdot q$ [resp. $k \cdot p \prec k \cdot q$].

There exist stably finite C*-algebras, even commutative C*-algebras with cancellation, which are not weakly unperforated:

Example 3.3.4. $C(\mathbf{T}^n)$ is not weakly unperforated for $n \geq 4$. $C(\mathbf{T}^4)$ has cancellation but is not weakly unperforated [**Bl 4**,6.10.2].

Weak perforation seems to have something to do with existence of minimal projections. The following fact, which is somewhat similar to 3.3.3, can be combined with subdivision arguments in the presence of arbitrarily small projections to prove weak unperforation in some cases (cf. 5.2.)

Proposition 3.3.5. [BH,III.2.6] Let A be a stably finite K-simple C*-algebra, p, q projections in $M_\infty(A)$. If $n\cdot p \precsim n\cdot q$ for some n, then $k\cdot p \precsim k\cdot q$ for all sufficiently large k.

3.4. In order to describe the final ingredient in the FCQ, we must digress to give a quick summary of the theory of ordered groups and K_0-theory. For a complete development of this theory see [**Bl 4**].

Recall that an *ordered group* is a pair (G,G_+) with G an abelian group and G_+ a subsemigroup satisfying $G_+ - G_+ = G$ and $G_+ \cap (-G_+) = \{0\}$. Write $y \leq x$ if $x-y \in G_+$. An element $u \in G_+$ is an *order unit* if for any $x \in G$ there is an $n>0$ with $x \leq nu$. G is *simple* if every nonzero element of G_+ is an order unit. A *scaled ordered group* is a triple (G,G_+,u), where (G,G_+) is an ordered group and u is an order unit. A *state* on a scaled ordered group (G,G_+,u) is a homomorphism $f:G \to \mathbf{R}$ with $f(G_+) \subseteq \mathbf{R}_+$ and $f(u)=1$. If $S(G)$ denotes the set of states on (G,G_+,u), then $S(G)$ is a compact convex set in the topology of pointwise convergence.

An abelian semigroup S is a *strict semigroup* if $x+y \neq x$ for all x, $y \in S$. Then the relation $<$ on S, defined by $y<x$ if there exists $z \in S$ with $x=y+z$, is a partial order. S is *archimedean* if for any x, $y \in S$ there is an n with $y<nx$. If S is a strict semigroup, G its Grothendieck group, and $\phi:S \to G$ the canonical homomorphism, set $G_+ = \phi(S) \cup \{0\}$; then (G,G_+) is an ordered group, which is simple if S is archimedean.

Definition 3.4.1. Let A be a (unital) C*-algebra. $V(A)$ is the semigroup of equivalence classes of projections in $M_\infty(A)$ with binary operation $[p]+[q]=[p \oplus q]$. $V_0(A)=V(A) \setminus [0]$. $K_0(A)$ is the Grothendieck group of $V(A)$ or of $V_0(A)$. If $\phi:V(A) \to K_0(A)$ is the canonical homomorphism, set $K_0(A)_+ = \phi(V(A))$.

Proposition 3.4.2. If A is stably finite, then $V_0(A)$ is a strict semigroup, which is archimedean if A is K-simple. So if A is simple and stably finite, $(K_0(A),K_0(A)_+,[1_A])$ is a simple scaled ordered group, called the *scaled ordered* K_0*-group* of A.

Recall that an ordered group is *weakly n-unperforated* if $nx>0$ implies $x>0$, and *weakly unperforated* if it is weakly n-unperforated for all n. A C*-algebra A is weakly n-unperforated if and only if its ordered K_0-group is weakly n-unperforated.

The importance of weak unperforation and the relevance to quasitraces and the FCQ comes from the following theorem of Goodearl and Handelman [**Bl 4**,6.8.5]:

Theorem 3.4.3. Let (G,G_+,u) be a weakly unperforated simple ordered group. Then G has the strict ordering from its states, i.e. if $x \in G$, then $x>0$ if and only if $f(x)>0$ for every state f on G.

Corollary 3.4.4. Let A be a stably finite K-simple C*-algebra which is strictly unperforated (i.e. is weakly unperforated and has strict cancellation.) If p and q are projections in matrix algebras over A, and $f([p]) < f([q])$ for every state f on the ordered K_0-group of A, then $p \precsim q$.

The only remaining step in the analysis of the FCQ is to relate the states on $K_0(A)$ with the normalized quasitraces on A. If $\tau \in QT(A)$, and p, q are projections in matrix algebras over A, then $\tau(p)=\tau(q)$ if $p \sim q$ and $\tau(p+q)=\tau(p)+\tau(q)$ if p and q are orthogonal. Thus τ defines a state f_τ on $K_0(A)$ by $f_\tau([p]-[q])=\tau(p)-\tau(q)$. In other words, there is a function $\chi: QT(A) \to S(K_0(A))$. It is easy to check that χ is a continuous affine function.

The function χ is in general far from injective; different traces can agree on all projections (this obviously happens in the commutative case, and can even occur for simple C*-algebras [**BK**].) The compact convex set $S(K_0(A))$ need not even be a Choquet simplex, at least in the non-simple case [**Bl 4**,6.10.4]. This leads to a question which, while unrelated to the FCQ, is of interest:

Question 3.4.5. If A is a simple C*-algebra, is $S(K_0(A))$ always a Choquet simplex?

There is no C*-algebra, simple or not, for which χ is known not to be surjective, i.e. there is no known state on a K_0-group which does not come from a trace. A counterexample might be difficult to construct, since there does not seem to be any way to get a handle on such a state (if they exist at all.) In contrast to the situation of cancellation and perforation, commutative C*-algebras are well behaved in this case: if A is commutative, then χ is always surjective [**Bl 4**,6.10.3].

Question 3.4.6. If A is a simple C*-algebra, is χ always surjective?

We will briefly return to the question of surjectivity of χ in section 6.

Actually, what we need for the FCQ (in addition to strict cancellation) is for $K_0(A)$ to have the strict ordering from the states which come from quasitraces. This is an apparently weaker condition than surjectivity of χ, but it is really the same thing due to the following fact.

Proposition 3.4.7. Let (G,G_+,u) be an ordered group, and K a closed convex subset of $S(G)$. If $f(x)>0$ for all $f \in K$ implies $nx>0$ for some $n>0$, then $K=S(G)$.

Proof: First suppose G is finitely generated. Then replacing G by the direct limit $G \xrightarrow{\phi_1} G \xrightarrow{\phi_2} G \xrightarrow{\phi_3} \cdots$, where ϕ_k is multiplication by k, we may assume $G=\mathbf{Q}^n \subseteq \mathbf{R}^n$, and every state on G extends uniquely to a linear functional on $\mathbf{R}^n$. If C_G is the closure of G_+ in $\mathbf{R}^n$, then C_G is a closed convex generating cone in $\mathbf{R}^n$. Similarly, let C_K and $C_{S(G)}$ be the nonnegative multiples of the functionals in K and $S(G)$ respectively; C_K and $C_{S(G)}$ are closed convex cones in $(\mathbf{R}^n)^*$. The assumption implies that C_G is the dual cone of C_K, and by definition $C_{S(G)}$ is the dual cone of C_G, so $C_K=C_{S(G)}$ by reflexivity and therefore $K=S(G)$.

Now examine a general G. If H is any finitely generated subgroup of G, then the restriction map $\rho_H: S(G) \to S(H)$ is surjective [**Bl 4**,6.8.3], and thus $\rho_H(K)=S(H)$ by the first part of the proof. So if f is any state on G, then there is a state $f_H \in K$ with $f|_H = f_H|_H$. As H runs over all finitely generated subgroups of G, $f_H \to f$. //

Corollary 3.4.8. If A is K-simple and stably finite, then $K_0(A)$ has the strict ordering from the states coming from quasitraces if and only if it is weakly unperforated and χ is surjective.

If χ is surjective, we say that A *has enough quasitraces.*

Summarizing, we have:

Theorem 3.4.9. Let A be a stably finite simple C*-algebra. Then A has stable strict comparability if and only if A is strictly unperforated (i.e. is weakly unperforated and has strict cancellation) and has enough quasitraces.

4. EXISTENCE OF SMALL PROJECTIONS

4.1. One reason simple C*-algebras may be better behaved with respect to the questions we are considering than, say, commutative C*-algebras is that an (infinite-dimensional) simple C*-algebra can be "subdivided" arbitrarily, in the sense that it is approximately a matrix algebra of arbitrarily large order over another simple C*-algebra. More precisely:

Proposition 4.1.1. Let A be an infinite-dimensional simple C*-algebra. Then, for any n, A contains two hereditary C*-subalgebras B and C, with $C \subseteq B$, such that B is naturally isomorphic to $M_n(C)$.

Proof: It suffices to assume $n=2$; the general case can then be obtained by iteration. A contains two nonzero orthogonal positive elements a and b. There is a y with $ayb \neq 0$ since A is simple [write $a = \sum_{i=1}^{n} y_i b z_i$; then $a y_i b z_i \neq 0$ for some i]; set $x = ayb$. Then x^*x and xx^* are orthogonal. Take C to be the hereditary C*-subalgebra of A generated by x^*x, B the hereditary C*-subalgebra generated by $x^*x + xx^*$. //

Note, however, that we cannot always choose B and C to be corners (subalgebras of the form pAp), so if we want to use this result in general we must deal with nonunital C*-algebras. To stay within our framework of only considering unital C*-algebras, we would like to be able to subdivide projections as in 4.1.1. And it will be important for technical arguments, and perhaps ultimately even for good results, to have "arbitrarily small projections;" our intuition on some problems seems better in this case too.

There are at least four natural senses in which we could say that a (simple) C*-algebra A has arbitrarily small projections:

(1) A contains a sequence of mutually orthogonal nonzero projections (equivalently, A contains a strictly increasing or decreasing sequence of projections)

(2) A contains no minimal projections

(3) For every nonzero projection p in $M_\infty(A)$ there is a nonzero projection q with $q \oplus q \precsim p$

(4) Every nonzero hereditary C*-subalgebra of A contains a nonzero projection.

(4)=>(3) by 4.1.1, and (3)=>(2)=>(1) is obvious. None of the reversed implications are known. If the underlying principles of the FCQ are true, they should all be the same.

Question 4.1.2. Which of the reverse implications are true?

Definition 4.1.3. A has the (DP) [resp. (SP)] property if A has arbitrarily small projections in sense (3) [resp. sense (4)].

4.2. Stable Rank.

It is appropriate at this point to digress to give a quick review of the concept of stable rank and its relationship with cancellation questions.

The theory of topological stable rank was developed by Rieffel [**Rf 1**] in analogy with the purely algebraic theory of Bass stable rank. The two notions were shown to coincide for C*-algebras by Herman and Vaserstein [**HV**]. Stable rank was originally viewed as a sort of "dimension theory" for C*-algebras, since it (essentially) reduces to ordinary dimension theory for commutative C*-algebras. However, stable rank behaves very differently for simple C*-algebras, and it is probably misleading to regard it as a dimension theory in this case. The underlying reason for the different behavior is the subdivision property described in 4.1.

Definition 4.2.1. Let A be a (unital) C*-algebra. Set

$$lg_n(A) = \{(x_1, \cdots, x_n) \in A^n \mid \sum_{i=1}^{n} x_i^* x_i \text{ is invertible}\}$$

The *stable rank* of A, denoted $sr(A)$, is the smallest n such that $lg_n(A)$ is dense in A^n. If $lg_n(A)$ is never dense, set $sr(A) = \infty$.

It is easy to see that $sr(A) = 1$ if and only if the invertible elements of A are dense in A. If p and q are projections in A, then an element y of qAp is said to be *left invertible* if there is an $x \in pAq$ with $xy = p$. If $sr(A) = n$, then the left invertible elements of $qM_\infty(A)p$ are dense, where $p = 1_A, q = 1_{M_n(A)}$.

Stable rank is important in studying cancellation. It is easy to show that if $sr(A) = 1$, then A has cancellation (4.3.6). In addition, for simple C*-algebras there is an important theorem due to Rieffel and Warfield [**Rf 2**] which uses a weakened version of the (DP) property:

Theorem 4.2.2. Let A be a simple C*-algebra. Suppose A contains a sequence (p_n) of projections such that

(1) For each n there is an r_n such that $2p_{n+1} \oplus r_n \precsim p_n \oplus r_n$

(2) There is a K such that $sr(p_nAp_n) \leq K$ for all n.

Then A has cancellation.

It is possible that condition (2) is automatic if A is stably finite. To look into this question, recall the behavior of stable rank with respect to forming matrix algebras: $sr(M_n(A))$ is roughly $sr(A)/n$, or more precisely

$$sr(M_n(A)) = \left\lceil \frac{sr(A)-1}{n} \right\rceil + 1$$

where $\lceil x \rceil$ denotes the least integer $\geq x$. So if $sr(A)=1$, then $sr(M_n(A))=1$, and if $1 < sr(A) < \infty$, then $sr(M_n(A))=2$ for sufficiently large n. This result suggests the following conjecture:

Conjecture 4.2.3. Let A be a C*-algebra, p a full projection in A. Then $sr(pAp) \geq sr(A)$. (More generally, if B is an arbitrary full hereditary C*-subalgebra of A, then $sr(B) \geq sr(A)$.)

The conjecture is true if A is an exact matrix algebra over pAp and should be true in general. A weak version is known [**Bl** 3,A6]: if $1 \precsim np$, then $sr(A) \leq sr(pAp) + n - 1$. This is enough to show that $sr(pAp) < \infty$ if and only if $sr(A) < \infty$.

If the conjecture is true, then if A is stably finite with (DP) there are four possibilities for the behavior of stable rank on the corners of A (or of matrix algebras over A). If p is a projection in $M_\infty(A)$, write $sr(p) = sr(pM_\infty(A)p)$.

(1) $sr(p) = 1$ for all p

(2) $sr(p) = 2$ for all p

(3) $sr(p) < \infty$ for all p; $sr(p) = 2$ for p large, $sr(p)$ unbounded for p small

(4) $sr(p) = \infty$ for all p.

It is not known whether cases (2), (3), or (4) can occur, i.e. there is no stably finite simple C*-algebra known to have stable rank greater than one. It seems unlikely that situation (3) could occur; if it did, there would be a canonical way of picking out a corner of "unit size" in the stable algebra of A, or more precisely a canonical way of normalizing the dimension functions on A. The intuition on simple C*-algebras not satisfying (DP) is less clear: if p is a minimal projection in A, then pAp is a natural candidate for a corner of "unit size." Simple C*-algebras in classes (3) or (4), if they exist, are candidates for counterexamples to 1.4.3 and the FCQ.

Let us collect together the open questions described above.

Question 4.2.4. Is there a (stably finite) simple C*-algebra A with $sr(A) = 2$? Is there one with (DP)?

Question 4.2.5. Is there a (stably finite) simple C*-algebra A with $2 < sr(A) < \infty$? Is there one with (DP)?

Question 4.2.6. Is there a stably finite simple C*-algebra A with $sr(A) = \infty$?

Recall that if A is simple and not stably finite, then $sr(A) = \infty$ [**Rf 1**,6.5].

There is a stable rank condition between (1) and (2) above. We say A has *stable rank* $< 1+\varepsilon$ if for any projections p, q in $M_\infty(A)$ with $p \not\precsim q$ the left-invertible elements of $qM_\infty(A)p$ are dense.

Question 4.2.7. Does every stably finite simple C*-algebra have stable rank $< 1+\varepsilon$?

4.3. Simple C*-algebras with (HP).

There is an existence of projections axiom stronger than (SP) which is extremely useful in attacking the FCQ (as well as in other contexts.) In this paragraph we will discuss this axiom and its consequences.

Definition 4.3.1. A C*-algebra A has (HP) if every hereditary C*-subalgebra of A has an approximate identity of projections. A has *stable (HP)* if $M_n(A)$ has (HP) for all n.

Question 4.3.2. Does (HP) imply stable (HP)?

There is an alternate way of phrasing (HP) which is sometimes useful, and which emphasizes the importance and naturality of the axiom.

Definition 4.3.3. An element x of a C*-algebra A is *well-supported* if there is a projection $p \in A$ with $x = xp$ and x^*x invertible in pAp.

x is well-supported if and only if either x^*x is invertible or 0 is an isolated point of the spectrum $\sigma(x^*x)$, i.e. if and only if $\sigma(x^*x) \subseteq \{0\} \cup [\varepsilon, \infty)$ for some $\varepsilon > 0$. So x is well-supported if and only if x^* is well-supported. Invertible elements and partial isometries are well-supported.

If x is well-supported, then x can be written as ua, where u is a partial isometry with $u^*u = p$ and a is an invertible positive element of pAp. $q = uu^*$ is a left support projection for x, and xx^* is invertible in qAq.

Proposition 4.3.4. If A has (HP) and p and q are projections of A, then the well-supported elements of qAp are dense in qAp.

Proof: Let $x \in qAp$. Let r be a projection in $\overline{x^*Ax}$ which is almost a unit for $(x^*x)^{1/2}$, and set $y = xr$. Then $y \in qAr \subseteq qAp$, and since $r \precsim n(x^*x)$ for some sufficiently large n, $y^*y = rx^*xr \geq (1/n)r$, so y^*y is invertible in rAr. Thus y is well-supported, and closely approximates x. //

Theorem 4.3.5. Let A be a C*-algebra. Then the following are equivalent:

(1) A has (HP)

(2) The well-supported self-adjoint elements of A are dense in A_{sa}.

(3) The invertible self-adjoint elements of A are dense in A_{sa}.

(4) The self-adjoint elements of A of finite spectrum are dense in A_{sa}.

Proof (outline): (1)=>(2) is almost identical to the proof of 4.3.4: if x is self-adjoint and r is a projection in xAx which is an approximate unit for x^*x, then rxr is a well-supported self-adjoint element closely approximating x.

(2)=>(3): if x is well-supported, then $x+\lambda 1$ is invertible for all sufficiently small nonzero λ.

(3)=>(4): suppose $x=x^*$ is given. We may assume $0 \le x \le 1$. Let $\{\lambda_1, \lambda_2, \cdots\}$ be the dyadic rationals in $[0,1]$. Set $x_1 = x$. For each n, let y_n be a well-supported self-adjoint element closely approximating $x_n - \lambda_n 1$, and set $x_{n+1} = y_n + \lambda_n 1$. Then x_n approximates x and its spectrum has gaps around $\lambda_1, \cdots, \lambda_n$; the approximate to x with finite spectrum can then be made from x_n by functional calculus.

(4)=>(1) is the trickiest part; see [**Pd**]. //

The notion of a well-supported element is closely connected with cancellation, and leads to a relationship between cancellation and stable rank in the (HP) case:

Proposition 4.3.6. If A is a C*-algebra, then A has cancellation if and only if the invertible elements of $M_n(A)$ are dense in the well-supported elements of $M_n(A)$ for all n.

Proof: Because of polar decomposition and the fact that invertible positive elements are dense in A_+, the invertible elements are dense in the well-supported elements if and only if every partial isometry can be arbitrarily approximated by invertibles. If A has cancellation and u is a partial isometry in $M_n(A)$, set $p = u^*u$, $q = uu^*$. By cancellation there is a partial isometry v with $v^*v = 1_{M_n(A)} - p$, $vv^* = 1_{M_n(A)} - q$; then $u + \varepsilon v$ is an invertible closely approximating u. Conversely, suppose $p \sim q$, and let u be a partial isometry in $M_n(A)$ with $u^*u = p$, $uu^* = q$. Approximate u closely by an invertible element x, and write $x = v(x^*x)^{1/2}$. Then v is unitary, and $vpv^* \approx v(x^*x)v^* = xx^* \approx q$, so p is unitarily equivalent to a projection close to q and hence also unitarily equivalent to q. //

Corollary 4.3.7. Let A be a C*-algebra with (HP). Then A has cancellation if and only if $sr(A) = 1$.

Proof: By 4.3.4 with $p = q = 1$ the well-supported elements are dense in A. To pass to matrix algebras use the fact that $sr(A) = 1$ implies $sr(M_n(A)) = 1$. //

(HP) also has an important bearing on the problem of existence of quasitraces:

Theorem 4.3.8. [BH] Let A be a stably finite simple C*-algebra with stable (HP). Then the map $\chi: QT(A) \to S(K_0(A))$ (3.4) is a bijection. So A has enough quasitraces.

While it is still an open question whether a stably finite simple C*-algebra with stable (HP) has stable strict comparability, we do have a weak comparability result which is of intrinsic interest and which can be used to reduce the FCQ in this case to another problem which may be more manageable.

Definition 4.3.9. Let A be a C*-algebra. A has *approximate comparability* if whenever p and q are projections in $M_\infty(A)$ and $\tau(p) < \tau(q)$ for all $\tau \in QT(A)$, then for any $\varepsilon > 0$ there is a projection $p' \leq p$ with $p' \prec q$ and $\tau(p-p') < \varepsilon$ for all $\tau \in QT(A)$.

Theorem 4.3.10. Let A be simple, with stable (HP), and with only finitely many extremal quasitraces. Then A has approximate comparability.

Proof: Let $M = l^\infty(A)/J_\omega$, where $J_\omega = \{(x_n) \mid \lim_\omega \tau(x_n^* x_n) = 0 \text{ for all } \tau\}$ be an ultraproduct of A. ($l^\infty(A)$ is the C*-algebra of all bounded sequences from A, and ω is a free ultrafilter on **N**.) Then M is a finite AW*-algebra [BH], a direct sum of factors with one for each extremal quasitrace. If p and q are projections in $M_k(A)$ with $\tau(p) \leq \tau(q)$ for all τ, then (regarding A as a subalgebra of M via constant sequences) there is a partial isometry $u \in M_k(M)$ with $u^*u = p$, $uu^* \leq q$. Represent u by a sequence $(u_n) \in l^\infty(M_k(A))$. We may assume $u_n \in qM_k(A)p$. Let x_n be a well-supported element of $qM_k(A)p$ with right support projection p_n and $\|u_n - x_n\| < 1/n$. Write $x_n = v_n a_n$ with v_n a partial isometry in $qM_k(A)p$; then the sequence (v_n) also represents u. Take $p' = p_n = v_n^* v_n$ for sufficiently large n. //

The assumption that A have only finitely many extremal quasitraces is almost certainly unnecessary; however, without this assumption M is not an AW*-algebra so the proof does not work.

Definition 4.3.11. A sequence of projections (p_n) in a C*-algebra A is a *null sequence* if the quasitraces approach 0 uniformly on the p_n, i.e. for any $\varepsilon > 0$ there is a k such that $\tau(p_n) < \varepsilon$ for all $n \geq k$ and for all $\tau \in QT(A)$.

(p_n) is a *thin sequence* if for every nonzero projection $q \in M_\infty(A)$, $p_n \precsim q$ for all sufficiently large n.

If A has strict comparability, then every null sequence is a thin sequence; if A has stable (HP) (or, more generally, (DP)), then every thin sequence is a null sequence.

Theorem 4.3.12. Let A be as in Theorem 4.3.10. Then the following are equivalent:

(1) Every null sequence in $M_\infty(A)$ is a thin sequence.

(2) A has stable strict comparability.

(3) A has cancellation and $K_0(A)$ is weakly unperforated.

(4) Every corner of A has stable rank ≤ 2 and $K_0(A)$ is weakly unperforated.

(5) A has stable rank 1 and $K_0(A)$ is weakly unperforated.

Proof: (5)=>(4) and (2)=>(1) are trivial; (3)=>(2) comes from 3.4.9 and 4.3.8, (4)=>(3) follows from 4.2.2, and (3)=>(5) is 4.3.6.

(1)=>(2): Suppose $\tau(p) < \tau(q)$ for all τ. Let r be a nonzero subprojection of q with $\tau(r) < \tau(q) - \tau(p)$ for all τ (this is possible by (HP).) By 4.3.10, choose $p_n \leq p$ with $p_n \prec q-r$ and $\tau(p-p_n) < 1/n$. Then by (1) $p-p_n \prec r$ for sufficiently large n, so $p = p_n + (p-p_n) \prec (q-r) + r = q$.

(2)=>(4): If A satisfies (2), so does every corner, so it suffices to show that $sr(A) \leq 2$. Let $x_1, x_2 \in A$; we must find $y_1, y_2 \in A$ arbitrarily close to x_1, x_2 with $y_1^* y_1 + y_2^* y_2$ invertible. Let $a = \begin{bmatrix} x_1 & 0 \\ x_2 & 0 \end{bmatrix} \in M_2(A)$, and approximate a closely by a well-supported matrix $b = \begin{bmatrix} z_1 & 0 \\ z_2 & 0 \end{bmatrix}$ with a nonzero left support projection $diag(p, 0)$. So $z_1^* z_1 + z_2^* z_2 = p$. For any τ, $\tau(diag(1-p, 0)) < 1 < \tau(1-q)$, where q is the right support projection of b. By (2) there is a partial isometry $u = \begin{bmatrix} w_1 & 0 \\ w_2 & 0 \end{bmatrix}$ with $u^*u = diag(1-p, 0)$, $uu^* \leq 1-q$. Set $y_i = z_i + \varepsilon w_i$ for small ε. //

5. EXAMPLES

In this section, we will discuss some classes of simple C*-algebras for which the answer to the FCQ is known or partially known, and also propose some good test algebras.

5.1. The first and most obvious class of examples of simple C*-algebras with strict comparability are the AF algebras. More generally, any simple C*-algebra which can be written as an inductive limit of (not necessarily simple) C*-algebras with stable strict comparability satisfies the FCQ. If X is a compact Hausdorff space for which $C(X)$ has cancellation and is weakly unperforated (e.g. S^n or $\mathbf{T}^n$ for $n \leq 3$), then $C(X) \otimes B$ has stable strict comparability for any finite-dimensional C*-algebra B, and hence inductive limits of such algebras satisfy the FCQ, even the FCQ2. So besides the simple AF algebras, algebras such as the Bunce-Deddens algebras [**Bl** 4,10.11.4] have strict cancellation. More generally, it seems to be a general principle that (almost) any simple C*-algebra which can be written in a nontrivial way as an inductive limit can be shown to have stable strict comparability. Some examples of this principle are given in 5.2 and 5.4.

5.2. Tensor Products With AF Algebras.

Using the results of 3.3, we can prove a fact which may be regarded as a "stable" version of the FCQ:

Theorem 5.2.1. Let B be an infinite-dimensional simple AF algebra. If A is a stably finite K-simple C*-algebra, then $A \otimes B$ has cancellation and is weakly unperforated. If A has enough quasitraces, then $A \otimes B$ has stable strict comparability.

Outline of proof: The last statement follows from the first statement and 3.4.9. The key observation in proving the first statement is that B has a Bratteli diagram in which the multiplicities of the partial embeddings become uniformly large far out in the diagram. We then use 3.3.3 to prove cancellation and 3.3.5 to prove weak unperforation. For more details see [**Bl 5**]. //

In particular, if X is connected and B is simple AF, then $C(X,B) = C(X) \otimes B$ has stable strict comparability ($C(X)$ has enough quasitraces by [**Bl 4**,6.10.3], and every quasitrace on $C(X)$ is a trace.) Since the class of C*-algebras with stable strict comparability is closed under inductive limits and finite direct sums, it follows that $C(X) \otimes B$ has stable strict comparability for all X, and hence so does any C*-algebra which is an inductive limit of such algebras.

I believe the same techniques can be used to prove the following generalization:

Conjecture 5.2.2. If A is a simple C*-algebra which is an inductive limit $\varinjlim C(X_i, B_i)$, where the X_i are compact and B_i finite-dimensional, then A satisfies FCQ2 and has cancellation.

One might also consider simple C*-algebras which are inductive limits of continuous fields of AF algebras, as in [**Bl 1**]. It is not clear that the same techniques will work for these algebras. Note that [**N**] contains a comparability theorem for certain such algebras proved with similar techniques.

5.3. Noncommutative Tori.

The noncommutative tori provide the oldest and probably best class (so far) of simple C*-algebras which satisfy the FCQ for nontrivial reasons. It is quite instructive to study these algebras in detail to appreciate how the topological obstructions to cancellation and unperforation, which at first seem to carry over to the noncommutative situation, actually disappear.

The two-dimensional noncommutative tori (irrational rotation algebras) have been studied by many authors, and have motivated some important work in K-theory such as the Pimsner-Voiculescu exact sequence for crossed products. The FCQ was settled for irrational rotation algebras by Rieffel [**Rf 2**], who has also obtained the definitive results in the higher-dimensional case [**Rf 3**] following work of others such as Elliott.

Definition 5.3.1. A *noncommutative n-torus* is a universal C*-algebra generated by n unitaries $\{u_1, \cdots, u_n\}$ which commute up to scalars, i.e. $u_j u_k = \lambda_{jk} u_k u_j$ for some $\lambda_{jk} \in \mathbf{C}$.

The λ_{jk} can be defined by a real bicharacter on $\mathbf{Z}^n$, a homomorphism $\theta : \mathbf{Z}^n \wedge \mathbf{Z}^n \to \mathbf{R}$ with $\lambda_{jk} = e^{2\pi i \theta(e_j \wedge e_k)}$, where $\{e_1, \cdots, e_n\}$ is the standard basis of $\mathbf{Z}^n$. We denote the noncommutative torus defined by the bicharacter θ by A_θ. (A_θ may also be regarded as the universal C*-algebra generated by a projective representation of $\mathbf{Z}^n$ with bicharacter θ.)

If the range of θ is contained in $\mathbf{Z}$, then $A_\theta = C(\mathbf{T}^n)$. More generally, if the range is contained in $\mathbf{Q}$, A_θ is stably isomorphic to $C(\mathbf{T}^n)$. In this case, we say θ is *rational*; otherwise θ is *irrational*. At the other extreme, if there is no nonzero $x \in \mathbf{Z}^n$ for which $\theta(x \wedge y) \in \mathbf{Q}$ for all y, then A_θ is simple; in this case θ is *totally irrational*.

If S is a subset of $\{u_1, \cdots, u_n\}$ with k elements, then the C*-subalgebra of A_θ generated by S is a noncommutative k-torus whose bicharacter is $\theta_S = \theta|_{S \wedge S}$. If θ is totally irrational, θ_S is not necessarily totally irrational; it can even be rational or integral. The inclusion of A_{θ_S} into A_θ induces an order-preserving homomorphism ϕ_S from the scaled ordered group $K_0(A_{\theta_S})$ to $K_0(A_\theta)$. More generally, this construction can be done for any subgroup S of $\mathbf{Z}^n$. General K-theory considerations show that ϕ_S is always injective.

An important question is whether ϕ_S is an order-isomorphism onto its image. If so, and if S were a subgroup of rank at least 4 on which θ_S is rational, then A_θ could not be weakly unperforated since $C(\mathbf{T}^4)$ is not weakly unperforated. At first it seemed that ϕ_S is always an order-isomorphism, so rationality of θ on four-dimensional subgroups appeared to be an obstruction to strict comparability. Rieffel proved that if θ is irrational on every rank 4 subgroup of $\mathbf{Z}^n$, then A_θ has strict comparability and even has cancellation.

I subsequently discovered an example where ϕ_S is not an order isomorphism: A_θ is the tensor product of $C(\mathbf{T}^4)$ with an irrational rotation algebra, with the subalgebra $C(\mathbf{T}^4) \otimes 1$. The argument is really just a special case of the one in 5.2. Rieffel was then able to upgrade his arguments to prove

Theorem 5.3.2. If θ is irrational, then A_θ has stable strict comparability and cancellation. In particular, if A_θ is simple, then it has stable strict comparability and cancellation. In addition, if A_θ is simple, it has a unique normalized quasitrace, which is a tracial state.

The proof of 5.3.2 uses 4.2.2 plus an explicit construction of projective modules over A_θ. The underlying reason why the result is true, and why the topological obstructions disappear, just as in 5.2, seems to be the idea of "infinite subdivisibility" of 4.1, along with 3.3.3 and 3.3.5, although none of these results explicitly enter into the proof.

5.4. Reduced Free Product C*-Algebras.

Perhaps the most intriguing class of examples studied so far are the reduced C*-algebras of free products of groups. While the results so far are not definitive, they are highly suggestive.

These are good test algebras for the FCQ, since in many respects their structure is pathological (e.g. they are not nuclear), in contrast to the simple C*-algebras considered in 5.1-5.3, which belong to the very nicest and most manageable class of non-type-I C*-algebras and which might be regarded as "too nice" to yield much insight into the general FCQ. In addition, the reduced C*-algebras of free product groups frequently have minimal projections, so they are quite different in this respect from the previous algebras, all of which have (DP).

Let G and H be nontrivial (discrete) groups, not both of order 2, and let $G * H$ be their free product. Then the reduced C*-algebra $C_r^*(G * H)$ is simple with unique trace [**PS**]. If G and H are amenable (and somewhat more generally), then the quotient map $\rho: C^*(G * H) \to C_r^*(G * H)$

induces an isomorphism on the K-groups [**Cu 5**].

Cuntz [**Cu 4**] calculated the K-theory of $C^*(G*H)$:

Theorem 5.4.1. $K_0(C^*(G*H))$ is naturally isomorphic to

$$(K_0(C^*(G)) \oplus K_0(C^*(H)))/<([1_{C^*(G)}],-[1_{C^*(H)}])>.$$

There are natural embeddings $\phi_G:C^*(G)\to C^*(G*H)$ and $\phi_H:C^*(H)\to C^*(G*H)$. 5.4.1 says that ϕ_{G*} and ϕ_{H*} are injective, the images together generate $K_0(C^*(G*H))$, and that there are no relations between the images other than the obvious one coming from $\phi_G(1_{C^*(G)})=\phi_H(1_{C^*(H)})$. If G and H are amenable, the same description works for $K_0(C_r^*(G*H))$.

What we would like is a description of the order structure on $K_0(C_r^*(G*H))$. The maps ϕ_{G*} and ϕ_{H*} are not generally order-isomorphisms onto their images, since the image must be simple. There is one obvious state f_τ on $K_0(C_r^*(G*H))$ coming from the unique trace τ, so the most obvious conjecture is the FCQ2:

Conjecture 5.4.2. $K_0(C_r^*(G*H))$ has the strict ordering from f_τ; if p and q are projections in matrix algebras over $C_r^*(G*H)$, then $p \prec q$ if and only if $\tau(p)<\tau(q)$.

Let us examine the case where G and H are finite cyclic groups, say $G=\mathbf{Z}_n$, $H=\mathbf{Z}_m$. $A=C_r^*(G*H)$ is generated by two finite-order unitaries u and v, and the images of the spectral projections of u and v generate $K_0(A)$. If $\{p_1,\cdots,p_n\}$ and $\{q_1,\cdots,q_m\}$ are the minimal spectral projections of u and v respectively, then $\tau(p_i)=1/n$, $\tau(q_j)=1/m$ for all i,j. If $r=lcm(n,m)$, then $f_\tau(x)$ is a multiple of $1/r$ for all $x\in K_0(A)$. It follows that A contains minimal projections; in fact if $m=n$ the p_i and q_j are minimal. The natural conjecture to examine is:

Conjecture 5.4.3. If $x\in K_0(A)$ and $f_\tau(x)=1/r$, then there is a projection $p\in A$ with $x=[p]$.

This conjecture would prove that $K_0(A)$ has a unique state and is unperforated, and would give some information about cancellation.

Comparability of projections in A was studied in [**ABH**], and conjecture 5.4.3 was partly confirmed. More specifically, if p and q are (not necessarily minimal) spectral projections of u and v respectively, and $\tau(p)<\tau(q)$, then $p \precsim q$. Many elements of $K_0(A)$ of trace $1/r$ were shown to come from projections in A, and the indications are quite strong that 5.4.3 is true. We have less firm evidence that A has cancellation.

The technique used in [**ABH**] is to look at the C*-subalgebra generated by p and q (as in the preceding paragraph.) This C*-algebra is a reduced free product of two copies of $\mathbf{C}^2$ with respect to certain traces, and its structure can be precisely described. However, while the technique does yield some fairly good information about the order structure of $K_0(A)$, it does not really penetrate deeply into the total internal structure of A since it only deals with what happens within the type I C*-subalgebras generated by two projections. A complete solution to 5.4.3 will probably require a more careful use of the structure of A as a simple C*-algebra.

There is a related case, however, that can be completely settled by the techniques of [**ABH**]. If G and H are infinite abelian torsion groups, then $C_r^*(G*H)$ can be written as an inductive limit of algebras of the above form. $C_r^*(G*H)$ will no longer have minimal projections; in fact, it will have enough small projections that one can use the results of [**ABH**] to show:

Theorem 5.4.4. Let G and H be infinite abelian torsion groups. If $x \in K_0(C_r^*(G*H))$ and $0 < f_\tau(x) < 1$, then there is a projection $p \in C_r^*(G*H)$ with $x = [p]$.

Turning to torsion-free groups, there is an excellent test algebra which will yield much insight into the question of whether K_0 of a stably finite simple C*-algebra can fail to be weakly unperforated. Consider $A = C_r^*(G*H)$, where $G = \mathbf{Z}^4$, $H = \mathbf{Z}$. The map $\phi_G : C^*(G) \cong C(\mathbf{T}^4) \to A$ induces an algebraic isomorphism $\phi_{G*} : K_0(C(\mathbf{T}^4)) \xrightarrow{\cong} K_0(A)$ which is order-preserving. Since $K_0(C(\mathbf{T}^4))$ is simple with a unique state, $K_0(A)$ (which of course is also simple) also has a unique state f_τ, and ϕ_{G*} is at least very nearly an order-isomorphism. In fact, $K_0(A)_+$ must be $\phi_{G*}(K_0(C(\mathbf{T}^4))_+)$ with none, some, or all of the perforation holes filled in. If the FCQ is true, all of the perforation must disappear.

Question 5.4.5. What is the positive cone of $K_0(A)$?

One may similarly test cancellation by considering $C_r^*(\mathbf{Z}^5 * \mathbf{Z})$.

It is very difficult to see how to approach these problems. Some of Voiculescu's recent work on reduced free products is relevant.

Another important question about these algebras is:

Question 5.4.6. What is the stable rank of $C_r^*(G*H)$?

It is quite possible that the stable rank is infinite for some or all of these algebras, providing answers to 4.2.6 and 4.2.7.

One could also consider other groups whose reduced C*-algebra is simple, such as HNN extensions. See 7.1 for some further comments.

6. COMPARABILITY FOR GENERAL POSITIVE ELEMENTS

In this section, we will examine the question of extending the notions of comparability of projections to general positive elements. It is important to have an understanding of this theory in the case of algebras without many projections, since in this case comparison theory for projections really doesn't tell much about the structure of the algebra. On the other hand, for an algebra with many projections (e.g. with stable (HP)), comparability theory for general elements can be reduced entirely to the theory for projections.

This section is more speculative than the previous ones, since the situation here is much less well understood. Not only are the technical problems much trickier, but intuition is also less clear and it is not even obvious what results to reasonably expect.

We will only deal with *simple* C*-algebras in this section, although some of the theory carries over to the nonsimple case. We will also continue to consider only unital algebras (but we will sometimes have to deal with hereditary C*-subalgebras which are nonunital.) Most of the theory here works for stably finite *algebraically simple* nonunital C*-algebras, but breaks down otherwise in the nonunital case. This theory is much more sensitive to stabilization than the theory for projections, due to the fact that there are positive elements in $A \otimes \mathbf{K}$ not equivalent to an element of $M_\infty(A)$.

Most of the basic theory in this case was developed by Cuntz [**Cu 2**]. There were also contributions by Handelman and the author [**BH**].

6.1. There are two aspects of "size" for a general positive element, "height" and "width" (size of support.) Since projections are uniformly of "unit height", only the "width" is important for them. We will concentrate only on the "width" in general. A good way of phrasing this is to think in terms of comparing hereditary C*-subalgebras instead of positive elements, although it will be convenient to actually express the relations in terms of elements.

Definition 6.1.1. If x is an element of a C*-algebra A, denote the hereditary C*-subalgebra of A generated by x^*x by A_x. A_x is the closure of x^*Ax.

A_x has a unit (is a corner) if and only if x is well-supported.

So for the purposes of comparison theory we will regard two positive elements as being the "same" if they generate the same (or unitarily equivalent) hereditary C*-subalgebras.

However, it is too restrictive to use unitary equivalence of the generated hereditary C*-subalgebras as our equivalence. Even in a UHF algebra it is extremely difficult to classify the hereditary C*-subalgebras up to unitary equivalence. One large part of the problem is that for any $\varepsilon > 0$ one can easily construct an x such that x^*x is orthogonal to a projection of trace $> 1-\varepsilon$ but xx^* is not orthogonal to any nonzero projection. It would make more sense, and turns out to yield a more tractable theory, to take an equivalence relation more similar to Murray-von Neumann equivalence by insisting that x^*x and xx^* be equivalent for all x.

The correct equivalence relation to use for a reasonable comparison theory is not obvious. There are two natural candidates:

Definition 6.1.2. Let a, b be positive elements of A.

$a \sim_s b$ if there is an $x \in A$ with $A_x = A_a$, $A_{x^*} = A_b$.

$a \sim b$ if there is a sequence (x_n) such that $(x_n^* x_n)$ is an approximate identity for A_a, $(x_n x_n^*)$ an approximate identity for A_b.

Since for any x we have $x^*x \sim_s xx^*$, we could extend the equivalence relations to arbitrary elements of A. But it seems to me to be simpler and more natural to only consider positive elements.

If $a \sim_s b$, then $a \sim b$; the converse is unclear. If a is well-supported and $a \sim b$, then b is well-supported and $a \sim_s b$. If a and b are projections, then $a \sim b$ in this sense if and only if a and b are Murray-von Neumann equivalent, so the notation is consistent. If a is well-supported with support projection p, then $a \sim p$.

If A has (HP), then $\sim$ and $\sim_s$ coincide, and every non-well-supported element is equivalent to one of the form $\sum \alpha_n p_n$, where the p_n are mutually orthogonal projections and $\alpha_n \to 0$.

Question 6.1.3. Are $\sim$ and $\sim_s$ always the same?

The relation $\sim_s$ is the most natural equivalence; however, for technical reasons we will from now on work only with $\sim$.

We would like to have a (reasonable) version of the FCQ for general positive elements. But in order to even phrase a statement, we need both a notion of comparability and numerical measures of size.

6.2. Let us try to develop a theory analogous to the construction of the ordered K_0-group.

Definition 6.2.1. Let $W(A)$ be the semigroup of $\sim$-equivalence classes of positive elements of $M_\infty(A)$ with $[a]+[b]=[a \oplus b]$.

There are two obvious subsemigroups of $W(A)$. The equivalence classes of well-supported elements form a subsemigroup which can be naturally identified with $V(A)$, and the classes of non-well-supported elements also form a subsemigroup which will be denoted $W_0(A)$. $V(A)+W_0(A) \subseteq W_0(A)$.

One interesting property that $W(A)$ has is that one can not only add finitely many elements, but one can also in certain cases add together infinitely many terms. More precisely, if $a_1, a_2, \cdots$ is a (bounded) sequence of mutually orthogonal elements of $M_n(A)$ (for a fixed n), then we can define $\sum_{k=1}^{\infty}[a_k] = [\sum_{k=1}^{\infty} 2^{-k} a_k]$. We may regard this process as the process of forming countable suprema of bounded sequences.

If A has stable (HP), then every element of $W_0(A)$ is a supremum of a strictly increasing sequence of projections (or a sum of a sequence of mutually orthogonal nonzero projections.) If (p_n) and (q_n) are strictly increasing, then $\sup\,[p_n] = [\sum 2^{-n}(p_n - p_{n-1})]$ $(p_0 = 0)$ is equivalent to $\sup\,[q_n]$ if and only if for each n there is an m with $p_n \precsim q_m$ and $q_n \precsim p_m$.

Example 6.2.2. Let A be a simple AF algebra, and set $\Delta = T(A)$. Each projection p defines a continuous strictly positive real-valued affine function $\hat{p}$ on Δ by $\hat{p}(\tau) = \tau(p)$. By stable strict comparability $p \prec q$ if and only if $\hat{p} < \hat{q}$ everywhere. If a is non-well-supported, we can regard $[a]$ as sup $[p_n]$ for a strictly increasing sequence p_n of projections. Write $\hat{a} = \sup \hat{p}_n$. If b is also non-well-supported, then $a \sim b$ if and only if $\hat{a} = \hat{b}$. $\hat{a}$ is a lower semicontinuous strictly positive affine function on Δ. Every continuous strictly positive affine function on Δ is a uniform limit of functions of the form $\hat{p}$ [**Bl 2**,3.1], and it follows easily that every strictly positive lower semicontinuous affine function on Δ occurs as $\hat{a}$ for some a. Thus $W_0(A) \approx Aff_l(\Delta)_{++}$, the additive semigroup of strictly positive lower semicontinuous affine functions on Δ.

Let us now return to the general case. We denote the Grothendieck group of $W(A)$ by $K_0^*(A)$. (This notation is unfortunate, since it appears to conflict with notational conventions of K-theory even though $K_0^*(A)$ has little to do with actual K-groups; but the notation is too well established to change.)

We will denote the class of a in $K_0^*(A)$ by $[[a]]$.

Proposition 6.2.3. If A is simple and not stably finite, then $K_0^*(A) = \{0\}$.

Proof: Suppose $M_k(A)$ is infinite. If a is any positive element of $M_n(A)$ for $n \geq k$, then there exists a bounded sequence (a_j) of mutually orthogonal elements of $M_n(A)$ all equivalent to a. Set $b = \sum_{j=2}^{\infty} 2^{-j} a_j$. Then $[a] + [b] = [a_1] + [b] = [a_1 + b] = [b]$ in $W(A)$. //

Question 6.2.4. If A is an infinite simple C*-algebra and a, b non-well-supported positive elements of A, is $a \sim b$?

This question is in effect asking whether every infinite simple C*-algebra is purely infinite with (HP).

6.3. Now let us assume A is simple and stably finite. Then $W(A)$ is a strict semigroup. $W(A)$ is never archimedean, since an element of $V(A)$ can never dominate an element of $W_0(A)$ in the algebraic ordering. Even $W_0(A)$ can fail to be archimedean: in example 6.2.2, $[a] < [b]$ in the algebraic ordering of 3.4 if and only if $\hat{a} < \hat{b}$ everywhere *and* $\hat{b} - \hat{a}$ is lower semicontinuous. So if Δ is infinite-dimensional $W_0(A)$ is not archimedean. $W(A)$ also rarely, if ever, has cancellation. For example, in 6.2.2 if p is a projection and a is a non-well-supported element with $\hat{a} = \hat{p}$, then for any non-well-supported b, $[a] + [b] = [p] + [b]$, but $[a] \neq [p]$.

Question 6.3.1. If A is simple and stably finite, does $W_0(A)$ have cancellation?

We could try to place an ordering on $K_0^*(A)$ as in the projection case, by taking the positive cone to be the image of $W(A)$. This does give an ordering, but it is not a very good one for the purposes of studying comparability. For example, in 6.2.2 we have $K_0^*(A)$ naturally isomorphic to $Aff_d(\Delta)$, the real-valued affine functions on Δ which can be written as differences of lower semicontinuous affine functions. The ordering from $W(A)$ gives a positive cone of $\{0\}\cup Aff_l(\Delta)_{++}$; it would be much better to have a positive cone of $Aff_d(\Delta)_+$, the nonnegative functions in $Aff_d(\Delta)$, or of $Aff_d(\Delta)_{++}\cup\{0\}$. So if $a \le b$ as positive elements, then $[[a]]$ is not in general $\le [[b]]$ in the ordering from $W(A)$.

There is also one additional phenomenon which complicates matters. In 6.2.2, if $a \le b$ as positive elements, then $\hat{a} \le \hat{b}$, but $\hat{a}$ is not in general strictly smaller that $\hat{b}$ (in fact we can have $\hat{a} = \hat{b}$ even if $a \neq b$.) This is in contrast to the case of projections: if $p < q$, then $\hat{p} < \hat{q}$ everywhere. So in the ordering on $K_0^*(A)$ we may want to use a stronger ordering on A_+ more in analogy with the strict ordering.

We define several notions of comparability to handle these problems. Some of the definitions are obvious, and others must be motivated later by the fact that they admit some comparability theorems.

Definition 6.3.2. Let a and b be positive elements in a C*-algebra A.
If $a \le b$, then $a \ll b$ if there is a $c \in A_b$ with $c \neq a$ and $ac = a$.
$a \precsim b$ if $a \sim a' \le b$; $a \prec b$ if $a \sim a' \ll b$.
$a \overline{\precsim} b$ if there is a sequence (x_n) with $x_n^* b x_n \to a$; $a \overline{\prec} b$ if $a \overline{\precsim} a' \ll b$.

It is obvious that $a \precsim b$ implies $a \overline{\precsim} b$, and $a \prec b$ implies $a \overline{\prec} b$. The converses are unclear. The great technical advantage in $\overline{\precsim}$ is that for fixed b, $\{a : a \overline{\precsim} b\}$ is closed.

Question 6.3.3. Are $\precsim$ and $\overline{\precsim}$ the same? Are $\prec$ and $\overline{\prec}$ the same?

The difficulty in proving this is very similar to the difficulty in 6.1.3. $\precsim$ and $\overline{\precsim}$ are the same for well-supported elements (in particular, for projections), as are $\prec$ and $\overline{\prec}$.

Note that if p and q are projections, then $p \ll q$ if and only if $p < q$, and $p \precsim q$ [resp. $p \prec q$] in this sense if and only if $p \precsim q$ [resp. $p \prec q$] in the usual sense. In the case of 6.2.2, $a \precsim b$ if and only if $\hat{a} \le \hat{b}$; $a \prec b$ if and only if $\hat{b} - \hat{a} \ge \varepsilon$ everywhere for some $\varepsilon > 0$.

Note that $\overline{\precsim}$ is called $\precsim$ in [**Cu 2**]; we have chosen to change notation in order to eventually get a cleaner statement of the FCQ with the same notation as for projections.

We are now ready to put two partial orderings on $K_0^*(A)$:

Definition 6.3.4. $K_0^*(A)_+ = \{[[b]]-[[a]] : a \overline{\precsim} b\}$; $K_0^*(A)_{++} = \{[[b]]-[[a]] : a \overline{\prec} b\}$.

These orderings give the right things in the case of 6.2.2: $K_0^*(A)_+ = Aff_d(\Delta)_+$, and $K_0^*(A)_{++}$ is the set of strictly positive elements of $Aff_d(\Delta)$ which are bounded away from 0 (this latter condition is not automatic for functions which are not lower semicontinuous.)

$(K_0^*(A), K_0^*(A)_{++})$ is always a simple ordered group; $(K_0^*(A), K_0^*(A)_+)$ is not in general simple. The two ordered groups are not so different: for example, they have the same state spaces. We will describe the state spaces in the next subsection.

6.4. We can use quasitraces to describe states on $K_0^*(A)$, but we must be careful: we cannot just set $f_\tau([[a]]) = \tau(a)$ since this function is not even well defined. Instead, we make the following definition:

Definition 6.4.1. If $\tau \in QT(A)$, define $d_\tau : M_\infty(A)_+ \to [0,\infty)$ by $d_\tau(a) = \lim_{n \to \infty} \tau(a^{1/n})$.

It is easy to show that d_τ is well defined, and that $d_\tau(a) \leq d_\tau(b)$ for $a \precsim b$ and $d_\tau(a+b) \doteq d_\tau(a) + d_\tau(b)$ for $a \perp b$. If p is a projection, then $d_\tau(p) = \tau(p)$.

Definition 6.4.2. A *dimension function* on A is a function $d : M_\infty(A) \to [0,\infty)$ satisfying

(1) $d(a) \leq d(b)$ if $a \precsim b$

(2) $d(a+b) = d(a) + d(b)$ if $a \perp b$.

d is *normalized* if $d(1_A) = 1$. Denote the set of normalized dimension functions on A by $DF(A)$.

$DF(A)$ is a compact convex set in the topology of pointwise convergence. The function $\delta : QT(A) \to DF(A)$ defined by $\delta(\tau) = d_\tau$ is an injective affine function which is not in general continuous. The next proposition summarizes important properties of δ:

Proposition 6.4.3. [**BH**] Let $\delta : QT(A) \to DF(A)$ be as above. Then

(1) The range of δ is the set of all lower semicontinuous dimension functions $LDF(A)$

(2) $LDF(A)$ is a face in $DF(A)$

(3) There is a (generally discontinuous) affine retraction $\rho : DF(A) \to LDF(A)$.

Proof of (3): Define $\rho(d)$ by $(\rho(d))(a) = \sup \{d(b) : b \ll a\}$. It is easy to check that $\rho(d)$ is a dimension function, and it is lower semicontinuous by [**BH**,I.1.5]. //

δ is rarely if ever surjective unless $QT(A)$ and $DF(A)$ are finite-dimensional. In fact, $DF(A)$ usually fails to be metrizable even if A is separable.

In all known cases, the range of δ is dense, and this is quite likely true in general. If so, there are nice consequences for comparability.

Question 6.4.4. Is $LDF(A)$ always dense in $DF(A)$, i.e. is the image of $\delta : QT(A) \to DF(A)$ always dense?

The next theorem may be regarded as the main result of [**Cu 2**], although it is almost immediate once the right definitions (which are really the main contribution of [**Cu 2**]) have been made.

Theorem 6.4.5. Let A be a stably finite simple C*-algebra. then the state space of $(K_0^*(A), K_0^*(A)_+)$ is naturally identified with $DF(A)$ under the correspondence $d \to f_d$, where

$f_d([[b]]-[[a]]) = d(b) - d(a)$.

Corollary 6.4.6. $DF(A)$ is also the state space of $(K_0^*(A), K_0^*(A)_{++})$. If $a, b \in M_\infty(A)_+$, and $d(a) < d(b)$ for all $d \in DF(A)$, then there is a $c \in M_\infty(A)_+$ and an n such that $n{\cdot}a \oplus c \precsim n{\cdot}b \oplus c$. If δ has dense range and $d_\tau(b) - d_\tau(a) \geq \varepsilon$ for some $\varepsilon > 0$ and for all $\tau \in QT(A)$, then there is a c and n with $n{\cdot}a \oplus c \precsim n{\cdot}b \oplus c$. [$n{\cdot}a$ means $a \oplus \cdots \oplus a$, not n times a in A.]

This result may be rephrased as follows, using 6.2.2 as motivation. Any nonzero positive a defines a continuous strictly positive real-valued affine function $\tilde{a}$ on $DF(A)$ by $\tilde{a}(d) = d(a)$. Composing $\tilde{a}$ with δ, a real-valued affine function $\hat{a}$ on $QT(A)$ is defined, which agrees with the previously defined $\hat{a}$ in the case of 6.2.2. Since δ is discontinuous, $\hat{a}$ is also discontinuous in general. If A has stable (HP), then $\hat{a}$ is always lower semicontinuous. We do have that $\hat{a} \geq \varepsilon$ everywhere for some $\varepsilon > 0$.

Question 6.4.6. Is $\hat{a}$ always lower semicontinuous for general A?

So 6.4.5 can be rephrased to say that if $\tilde{a} < \tilde{b}$ everywhere, then $n{\cdot}a \oplus c \precsim n{\cdot}b \oplus c$ for some n and c, and so the same is true if δ has dense range and $\hat{b} - \hat{a} \geq \varepsilon$.

We can now phrase a version of the Fundamental Comparability Question for general positive elements:

6.4.7. Fundamental Comparability Question, Version 4 (FCQ4). Let A be a stably finite simple C*-algebra, a, b positive elements of $M_\infty(A)$. If $\hat{b} - \hat{a} \geq \varepsilon$ everywhere (i.e. $d_\tau(b) - d_\tau(a) \geq \varepsilon$ for all $\tau \in QT(A)$) for some $\varepsilon > 0$, is $a \prec b$?

Of course, the FCQ4 includes the FCQ3 as a special case.

6.5. We can analyze the constituent parts of the FCQ4 just as we did for the FCQ3. The components are strict cancellation, weak unperforation, and existence of enough quasitraces. In addition, there is the question of the agreement of $\prec$ and $\precsim$.

Weak unperforation is straightforward to define in this case, and existence of enough quasitraces is simply the question of whether the range of δ is dense. Strict cancellation can be phrased as follows:

Definition 6.5.1. $W(A)$ has *strict cancellation* if $a \oplus c \prec b \oplus c$ implies $a \prec b$.

Theorem 6.5.2. Let A be a stably finite simple C*-algebra for which the relations $\prec$ and $\precsim$ agree. Then A satisfies FCQ4 if and only if $(K_0^*(A), K_0^*(A)_{++})$ is weakly unperforated, $W(A)$ has strict cancellation, and the range of $\delta : QT(A) \to DF(A)$ is dense.

Strict cancellation is tricky business since $W(A)$ does not have cancellation. There is a weaker version of the question which would be of interest:

Question 6.5.3. Let A be a stably finite simple C*-algebra. If $a \oplus c \prec b \oplus c$ for some c, is $a \oplus d \prec b \oplus d$ for all non-well-supported $d \succsim c$?

If so, the theory of ordered semigroups [**Bl 5**] can be applied to the study of $W_0(A)$ to obtain results such as those of 5.2.

The basic question in the FCQ4 which is new is whether there are "different ways" a positive element can fail to be well-supported, i.e. whether there are different ways the element can approach zero at the "edge" of its support (and if so whether the equivalence relation ~ is flexible enough to deal with the differences.) In commutative C*-algebras, especially with disconnected base space, it is hopelessly complicated to describe the behavior of a function near the edge of its support; however, examples such as 6.2.2 suggest the possibilities are very much more limited in the simple case, and in fact the comparability properties of an element in a simple C*-algebra may very well only depend on the numerical size of the support (as measured by the dimension functions) and whether or not the element is well-supported.

In fact, for non-well-supported elements an even stronger statement than the FCQ4 may be true:

6.5.4. Fundamental Comparability Question, Version 5 (FCQ5). Let A be a stably finite simple C*-algebra, a, b positive elements of $M_\infty(A)$ which are not well-supported. If $d_\tau(a) \leq d_\tau(b)$ for all $\tau \in QT(A)$, is $a \precsim b$?

FCQ5 => FCQ4 as follows. It suffices to prove FCQ5 => FCQ3. If $\hat{q} - \hat{p} \geq \varepsilon$, then we can find a non-well-supported c with $\hat{c} < \varepsilon/2$ everywhere by 4.1.1, and it is easy to construct a non-well-supported $b \leq q$ with $\hat{q} - \hat{b} < \varepsilon/2$ everywhere by a similar argument. Then $p \oplus c$ and b satisfy the hypotheses of the FCQ5.

In the presence of stable (HP), as one might expect, FCQ3, FCQ4, and FCQ5 are all equivalent, and are equivalent to the conditions of 4.3.12. If in addition the algebra has a unique quasitrace, these conditions are also equivalent to *factoriality* as defined in [**Cu 1**].

Question 6.5.5. Let A be a simple C*-algebra with unique quasitrace. Is A factorial?

If A has a unique quasitrace and projections of arbitrarily small trace, and satisfies FCQ5, then every non-well-supported element is equivalent to an element of the form $\sum \alpha_n p_n$ for a sequence p_n of mutually orthogonal projections and $\alpha_n \to 0$. This in turn implies that A has stable (HP). So good test algebras for FCQ5 are the irrational rotation algebras, which are not known to have (HP).

In connection with the question of uniqueness of behavior at the "edge" of the support, the following question is of interest. If the answer is yes, then $K_0^*(A)$ can be constructed and analyzed entirely using the (more well-behaved) semigroup $W_0(A)$ instead of $W(A)$.

Question 6.5.6. Let A be a stably finite simple C*-algebra, p a projection in $M_\infty(A)$. Is there a non-well-supported a such that $[[p]] = [[a]]$, i.e. are there non-well-supported elements a and b such that $p \oplus b \sim a \oplus b$?

6.6. There is a construction which can be done for simple C*-algebras which is based on the ideas of 4.3.11 and which seems to be quite useful in studying comparability.

Definition 6.6.1. A sequence (a_n) of positive elements of a C*-algebra A is called a *null sequence* if $\hat{a}_n \to 0$ uniformly, i.e. for any $\varepsilon > 0$ there is a k such that $d_\tau(a_n) < \varepsilon$ for all $n \geq k$ and for all $\tau \in QT(A)$.

(a_n) is a *thin sequence* if for any nonzero $b \in A_+$, $a_n \precsim b$ for all sufficiently large n.

More generally, if ω is a free ultrafilter on N, (a_n) is *null along* ω if for any $\varepsilon > 0$ $\{n : d_\tau(a_n) < \varepsilon$ for all $\tau \in QT(A)\}$ belongs to ω. (a_n) is *thin along* ω if for any nonzero b $\{n : a_n \precsim b\}$ is in ω.

The set of bounded null sequences forms an ideal in $l^\infty(A)$, as does the set of bounded sequences null along ω for any ω, because of the subadditivity property of dimension functions. The closure of this ideal is $J_\omega = \{(a_n) : \lim_\omega \tau(a_n) = 0 \text{ uniformly}\}$. J_ω is the sum of the ideal of ω-null sequences and the ideal $c_0(A)$ of sequences converging to 0 in norm.

The set of thin sequences does not in general form an ideal. However, if A is infinite-dimensional and simple, then 4.1.1 can be used to prove that the thin sequences do form an ideal:

Proposition 6.6.2. Let A be an infinite-dimensional C*-algebra. Then, for any ω, the set of bounded sequences which are thin along ω form an ideal T_ω in $l^\infty(A)$.

Proof: The difficulty is in proving that T_ω is closed under addition. If (a_n) and (b_n) are sequences thin along ω, and c is a nonzero positive element, let c_1 and c_2 be equivalent orthogonal elements of A_c. Let $S_1 = \{n : a_n \precsim c_1\}$, $S_2 = \{n : b_n \precsim c_2\}$. Then $S_1, S_2 \in \omega$, so $S = S_1 \cap S_2$ is also in ω. If $n \in S$, then $a_n + b_n \precsim a_n \oplus b_n \precsim c_1 \oplus c_2 \precsim c$. //

Of course, every thin sequence is a null sequence. The converse is very close to the FCQ4.

Question 6.6.3. Is every null sequence a thin sequence? Is the closure of T_ω equal to J_ω?

If A is purely infinite with (SP), then every bounded sequence is a thin sequence; otherwise, no constant sequence is a thin sequence. In this case, the inclusion of A into $l_\infty(A)$ via constant sequences drops to an inclusion of A into $l^\infty(A)/\overline{T}_\omega$. It is a promising approach to the FCQ to use this quotient algebra. For example, it may be finite if and only if $\overline{T}_\omega = J_\omega$. Also, the quotient is very similar to an AW*-algebra, so it may be finite if and only if it is stably finite; this would give a method of attacking problem 1.4.1.

One can more generally take a sequence A_n of simple C*-algebras containing A and do a similar construction. For example, we could take $A_n = M_n(A)$ with A embedded as diagonal matrices. This construction also seems to be relevant to 1.4.1. I plan to pursue this approach in the future.

6.7. Finally, we examine what this theory has to say about the theory of comparability of projections. If p, q are projections and $p \oplus a \sim q \oplus a$ for some non-well-supported a, it does not follow that $p \sim q$ or even that $p \oplus r \sim q \oplus r$ for some projection r. 6.2.2 provides a counterexample if $\hat{p} = \hat{q}$ but $p \not\sim q$. This suggests:

Question 6.7.1. If $p \oplus a \precsim q \oplus a$ for some non-well-supported a, is $p \precsim q$? Is $p \oplus r \precsim q \oplus r$ for some projection r?

The FCQ4, of course, implies the first question.

There is another way of looking at the situation. There is an order-preserving homomorphism $\phi : (K_0(A), K_0(A)_+) \to (K_0^*(A), K_0^*(A)_{++})$ defined by $\phi([p]) = [[p]]$. (ϕ is induced by the inclusion of $V(A)$ into $W(A)$.) ϕ is not one-to-one in general, even if A is AF; in the situation of 6.2.2 $\phi([p]) = \phi([q])$ if and only if $\hat{p} = \hat{q}$.

Question 6.7.2. If $x \in K_0(A)$ and $\phi(x) > 0$ (i.e. $\phi(x) \in K_0^*(A)_{++}$), is $x > 0$?

There is a continuous affine map $\sigma : DF(A) = S(K_0^*(A)) \to S(K_0(A))$ given by restriction; $\sigma \circ \delta$ is the map χ of 3.4. The retraction ρ of 6.4.3 shows that the range of σ is the same as the range of χ (if $d \in DF(A)$, then $(\rho(d))(p) = d(p)$ for any projection p.) So question 6.7.2 is closely related to the question of whether χ is surjective, as well as whether any possible perforation in $K_0(A)$ disappears in $K_0^*(A)$.

7. SOME CONSEQUENCES

In this section, we will discuss two questions which seem to be only vaguely related to the FCQ, and show that there is in fact quite a strong connection.

7.1. The first question is a nonstable K_1 question. Recall that if A is a (unital) C*-algebra, and $U_n(A)$ denotes the unitary group of A, then $K_1(A)$ is the direct limit of $U_n(A)/U_n(A)_0$ under the embeddings $u \to u \oplus 1$. Thus there is a natural homomorphism $\mu_n : U_n(A)/U_n(A)_0 \to K_1(A)$.

μ_n is not generally either injective or surjective; in fact, $U_n(A)/U_n(A)_0$ is not generally abelian, although $K_1(A)$ is always abelian. Examples are given in [**Bl** 4,§9]. However, none of these examples are simple C*-algebras, so we may ask:

Question 7.1.1. If A is a simple C*-algebra, is the map $\mu_1 : U_1(A)/U_1(A)_0 \to K_1(A)$ always an isomorphism?

It might seem unreasonable to hope for a positive answer, but there are enough partial results to suggest that 7.1.1 may be true.

The best theorem so far on this question is the following result due to Rieffel and Vaserstein [**Rf 1**,10.12]:

Theorem 7.1.2. Let A be a (unital) C*-algebra. If $n \geq sr(A)$ then μ_n is surjective. If $n \geq sr(A)+2$, then μ_n is also injective.

There are non-simple examples to show that this result is the best possible in general. However, this plus the subdivision and stable rank ideas of section 4 suggest that one can do much better in the simple case. For example, the following corollary is immediate, and is applicable to a wide class of examples.

Corollary 7.1.3. Let A be a simple C*-algebra. If for some n (usually $n=1$ or 2) A contains $n+2$ equivalent orthogonal projections $p_1, \cdots, p_{n+2}$ with $sr(p_iAp_i) \leq n$, then μ_1 is an isomorphism. In particular, if $sr(A)=1$ and A contains three equivalent orthogonal projections, then μ_1 is an isomorphism.

So 7.1.1 is related to question 4.2.5. The condition on existence of projections can be relaxed; in fact, using 4.1.1 and 7.1.2 one can prove

Corollary 7.1.4. If A is a simple C*-algebra with stable rank 1, then μ_1 is an isomorphism.

The direct connection with the FCQ comes from the following construction, which shows that unitaries may be moved homotopically into "small" hereditary C*-subalgebras. We would like to answer the following question:

Question 7.1.5. Let A be a simple C*-algebra. If u is a unitary and b is a nonzero positive element in A, then is u homotopic to a unitary of the form $1+x$ for some $x \in A_b$?

This would get us close to 7.1.1: it would show that μ_1 is always surjective, and would also show that $U_1(A)/U_1(A)_0$ is always abelian.

We first need some notation.

Definition 7.1.6. Let $0 \leq \alpha < \beta \leq 1$. $f_{\alpha,\beta}$ is the continuous function from $[0,1]$ to $[0,1]$ which is 0 on $[0,\alpha]$, linear on $[\alpha,\beta]$, and 1 on $[\beta,1]$. $g_{\alpha,\beta}$ is the continuous function from S^1 to S^1 defined by $g_{\alpha,\beta}(e^{2\pi i t}) = e^{2\pi i f_{\alpha,\beta}(t)}$.

Now suppose u is a unitary in A. Since $g_{\alpha,\beta}$ is homotopic to the identity map on S^1, the unitary $g_{\alpha,\beta}(u)$ is homotopic to u. And $g_{\alpha,\beta}(u)$ is of the form $1+x$, where x belongs to the hereditary C*-subalgebra of A generated by the functions of u which are supported on the interval from $e^{2\pi i\alpha}$ to $e^{2\pi i\beta}$. If α and β are suitably chosen and close together, this hereditary C*-subalgebra should be "small."

Suppose A is purely infinite with (SP), i.e. every nonzero hereditary C*-subalgebra contains an infinite projection. Then there is an infinite projection orthogonal to this hereditary C*-subalgebra, i.e. there is a projection p with $1-p$ infinite and u homotopic to a unitary of the form $1+x$, $x \in A_p$. It follows easily that 7.1.5 is true for such an A, and in fact μ_1 is a bijection. This argument is due to Cuntz [**Cu 3**,1.9].

Theorem 7.1.7. If A is a purely infinite simple C*-algebra with (SP), then $\mu_1 : U_1(A)/U_1(A)_0 \to K_1(A)$ is an isomorphism.

Thus 7.1.1 is true in the two extreme cases where A is purely infinite or $sr(A)=1$ (under a mild hypothesis on existence of projections, which may be unnecessary.)

Now suppose A is stably finite. Choose a sequence of disjoint intervals $[\alpha_n,\beta_n]$ of $[0,1]$, and let $u_n = 1+x_n$ be $g_{\alpha_n,\beta_n}(u)$; set $a_n = x_n^* x_n$. Then (A_{a_n}) is a sequence of mutually orthogonal hereditary C*-subalgebras of A, and u is homotopic to a unitary in $1+A_{a_n}$ for each n. For any $\tau \in QT(A)$ we have $\sum d_\tau(a_n) \le 1$, so the functions $\hat{a}_n$ approach 0 pointwise on $QT(A)$ (and in fact somewhat more strongly.)

Unfortunately, a sequence like this need not converge uniformly to 0; a counterexample can easily be constructed in the case of 6.2.2 if Δ is infinite-dimensional. It is not clear in general that the $[\alpha_n,\beta_n]$ can be chosen to make the sequence $\hat{a}_n$ converge uniformly to 0.

But if A has only finitely many extremal quasitraces, then the $\hat{a}_n$ must converge uniformly to 0. If A satisfies the FCQ4, then for any b we can find an n so that $a_n \precsim b$. In fact, by a more delicate argument we can show that eventually a_n is *unitarily* equivalent to an element of A_b via a unitary in $U_1(A)_0$. Thus we can show that FCQ4 + (finitely many extremal quasitraces) implies 7.1.5 and therefore (almost) 7.1.1.

I actually think there is stronger reason to believe 7.1.1 has a positive answer than the FCQ. There is a great deal of freedom in homotopically moving a unitary; the way we have done it here using functional calculus is extremely special, and the fact that we can get close to the result by these rather trivial methods is great cause for optimism.

7.2. Another question which is closely related to 7.1.1 and fairly closely related to the FCQ is the agreement of the different equivalence relations on projections. The three are Murray-von Neumann equivalence, unitary equivalence, and homotopy. These are successively stronger relations which agree on $M_\infty(A)$ but not in general on A. For a complete account of the story see [**Bl 4**,§4].

Question 7.2.1. If p and q are nontrivial projections in a simple C*-algebra A, with $p \sim q$, are p

and q homotopic?

The answer is yes if A is purely infinite [**Cu 3**]: the reason is basically that in this case $1-p$ and $1-q$ contain three copies of 1, and $p \sim q$ implies that $p \oplus 0 \oplus 0 \oplus 0$ and $q \oplus 0 \oplus 0 \oplus 0$ are homotopic.

If A is stably finite, the question of whether equivalence implies unitary equivalence is precisely the cancellation question (3.2.4.) If p and q are unitarily equivalent via a unitary u, then they are homotopic precisely if and only if there is a unitary v homotopic to u which commutes with p. So it would suffice to find a unitary of the form $1+x$, with $x \in A_p$, which is homotopic to u (cf. 7.1.5.) Therefore, if 7.1.1 is true then 7.2.1 is also true.

I believe 7.2.1 is very likely true.

REFERENCES

[**ABH**] J. Anderson, B. Blackadar, U. Haagerup, Minimal projections in Choi's algebra, to appear.

[**Bl 1**] B. Blackadar, A simple unital projectionless C*-algebra, *J. Operator Theory* **5** (1981), 63-71.

[**Bl 2**] B. Blackadar, Traces on simple AF C*-algebras, *J. Functional Anal.* **38** (1980), 156-168.

[**Bl 3**] B. Blackadar, A stable cancellation theorem for simple C*-algebras, *Proc. London Math. Soc.* **47** (1983), 303-305.

[**Bl 4**] B. Blackadar, *K-Theory for Operator Algebras*. MSRI Publication Series, Springer-Verlag, New-York/Heidelberg/Berlin/Tokyo, 1986.

[**Bl 5**] B. Blackadar, Rational C*-algebras and nonstable *K*-theory, to appear.

[**BH**] B. Blackadar and D. Handelman, Dimension functions and traces on C*-algebras, *J. Functional Anal.* **45** (1982), 297-340.

[**BK**] B. Blackadar and A. Kumjian, Skew products of relations and the structure of simple C*-algebras, *Math. Z.* **189** (1985), 55-63.

[**Cu 1**] J. Cuntz, The structure of addition and multiplication in simple C*-algebras, *Math. Scand.* **40** (1977), 215-233.

[**Cu 2**] J. Cuntz, Dimension functions on simple C*-algebras, *Math. Ann.* **233** (1978), 145-153.

[Cu 3] J. Cuntz, *K*-Theory for certain C*-algebras, *Ann. of Math.* **113** (1981), 181-197.

[Cu 4] J. Cuntz, The *K*-groups for free products of C*-algebras, *Operator Algebras and Applications* (ed. R. V. Kadison), Proc. Symp. Pure Math. 38, Amer. Math. Soc., Providence, 1981, pt. 1, 81-84.

[Cu 5] J. Cuntz, *K*-Theoretic amenability for discrete groups, *J. Reine Angew. Math.* **344** (1983), 180-195.

[CP] J. Cuntz and G. Pedersen, Equivalence and traces on C*-algebras, *J. Funct. Anal.* **33** (1979), 135-164.

[Hd] D. Handelman, Homomorphisms of C* algebras to finite AW* algebras, *Michigan Math. J.* **28** (1981), 229-240.

[HV] R. Herman and L. Vaserstein, The stable range of C*-algebras, *Invent. Math.* **77** (1984), 553-555.

[MvN] F. Murray and J. von Neumann, On rings of operators, *Ann. of Math.* **37** (1936), 116-229.

[N] V. Nistor, On the homotopy groups of the automorphism group of an AF-C*-algebra, to appear.

[PS] W. Paschke and N. Salinas, C*-Algebras associated with free products of groups, *Pacific J. Math.* **82** (1979), 211-221.

[Pd] G. Pedersen, The linear span of projections in simple C*-algebras, *J. Operator Theory* **4** (1980), 289-296.

[Rf 1] M. Rieffel, Dimension and stable rank in the *K*-theory of C*-algebras, *Proc. London Math. Soc.* (3) **46** (1983), 301-333.

[Rf 2] M. Rieffel, The cancellation theorem for projective modules over irrational rotation algebras, *Proc. London Math. Soc.* (3) **47** (1983), 285-302.

[Rf 3] M. Rieffel, Projective modules over higher dimensional non-commutative tori, to appear.

[W] J. Wright, On AW*-algebras of finite type, *J. London Math. Soc.* **(2) 12** (1976), 431-439.

Interpolation for Multipliers

Lawrence G. Brown,
Dept. of Mathematics, Purdue University, West Lafayette,
Indiana 47907,U.S.A.

In this talk we state an existence theorem for multipliers and outline its proof. We have some applications of this theorem. Since techniques for construction of multipliers seem very interesting, we hope that more applications of the theorem or its proof will be found. We also hope that others will try to improve on the statement or proof of the theorem.

Consider a C^*-algebra A and its enveloping W^*-algebra A^{**}. There are several concepts of semicontinuity for elements of A^{**}_{sa} (see [2]). We will use two of these in the talk. For $h \in A^{**}_{sa}$, h is called *strongly lsc* if $h = \lim h_n$ (norm limit), where each h_n is the limit of a monotone increasing net of elements of A. h is called q-lsc if the spectral projection $E_{(-\infty,t]}(h)$ is a closed projection for each t in $\mathbb{R}$.

Theorem 0. If $h,k \in A^{**}_{sa}$, he is strongly lsc, k is strongly usc, and $h \geq k$, then there is a in A such that $h \geq a \geq k$.

q

Theorem 1. If A is σ-unital, $h,k \in A^{**}_{sa}$, h is q-lsc, k is q-usc, and $h \geq k$, then there is x in $M(A)_{sa}$ such that $h \geq x \geq k$.

Theorem 0 is the natural theorem on "strong interpolation" and is stated only for background. Theorem 1, the subject of the talk, deals with "middle interpolation." In the special case where A is unital, $M(A) = A$, but that hypothesis of Theorem 1 is stronger than that of Theorem 0. Both theorems are non-commutative analogues of the following theorem of general topology: If X is a normal topological space, f and g are real valued functions on X, f is lsc, g is usc, and $f \geq g$, then there is a continuous function h on X such that $f \geq h \geq g$. For complete proofs of both theorems, as well as further background and motivation, see [3].

Before proceeding we give two further definitions involving the prefix "q-". For h,k self-adjoint we say that $h \overset{q}{\geq} k$ if $E_{(-\infty,s]}(h)$ is orthogonal to $E_{[t,\infty)}(k)$ whenever $s < t$. If h and k commute, $h \overset{q}{\geq} k$ if and only if $h \geq k$. In general, $h \overset{q}{\geq} k$ implies $h \geq k$. If B and C are hereditary C^*-subalgebras of A with open projections p and q, we say that B and C *q-commute* if p and q commute.

Lemma 1. Let B and C be q-commuting hereditary C^*-subalgebras of A. Then there is an increasing approximate identity (r_i) of $B \cap C$ such that $\|b(1-r_i)c\| \to 0$ for all b in B, c in C.

The proof of this is similar to that of 3.12.14 of [4] (quasi-central approximate identities).

Lemma 2. Let A be a σ-unital C^*-algebra and p,q closed projections in A^{**}dd such that $pq = 0$. Then there is h in M(A) such that $p \leq h \leq 1 - q$.

Remark. This is a non-commutative analogue of Urysohn's lemma, appropriate to the "middle" concept of semicontinuity. The "strong" version of Urysohn's lemma was proved by Akemann [1].

Sketch of proof. Let B,C be the hereditary C^*-subalgebras whose open projections are $1-p$, $1-q$. Let b,c,d be strictly positive elements of $B, C, B \cap C$. Choose $\epsilon_n > 0$ such that $\epsilon_n \searrow 0$ $(n \geq 1)$ and $\sum \epsilon_n < \infty$. We construct recursively elements x_n, x_n' of $(B \cap C)_+$ $(x_1' = 0)$ such that:

(1) $s_n = \sum_1^n (x_k + x_k') \leq 1$

(2) $\|bx_n\|$, $\|x_n' c\| \leq 2\epsilon_{n-1}^{\frac{1}{2}}$, $n > 1$

(3) $\|b(1 - s_n)c\| < \epsilon_n$

(4) $\|(1 - s_n)d\| < 2^{-n}$

We let $h = p + \sum_1^\infty x_k$. ((1) shows that this sum makes sense in A^{**}.)

From (4) (s_n) is an approximate identity of $B \cap C$ and hence $1 - h = q + \sum_1^\infty x_k'$. Then (2) implies that bh and $(1 - h)c$ are in A. Thus $h \in M(A)$.

The construction of x_k, x_k' for $k = 1,2$ will suffice to explain the recursive construction. First apply Lemma 1 to obtain $S_1 = x_1$ in $B \cap C$ satisfying (1), (3), (4). Then apply Lemma 1 again with b replaced by $v(1-s_1)^{\frac{1}{2}}$ and c replaced by $(1 - s_1)^{\frac{1}{2}}c$ to obtain z in $B \cap C$ such that $0 \le z \le 1$,
$\|(1 - z)(1 - s_1)^{\frac{1}{2}}d\| < 2^{-2}$, and $\|b(1 - s_1)^{\frac{1}{2}}(1 - z)(1 - s_1)^{\frac{1}{2}}c\| < \varepsilon_2$.
Then $\|b(1-s_1)^{\frac{1}{2}}z(1 - s_1)^{\frac{1}{2}}c\| < \varepsilon_2 \le 2\varepsilon_1$. In other words $e_1 e_2$ is small, where $e_1 = |b(1 - s_1)^{\frac{1}{2}}z^{\frac{1}{2}}|$ and $e_2 = |c(1 - s_1)^{\frac{1}{2}}z^{\frac{1}{2}}|$. Using the continuous functional calculus, we can find y in $B \cap C$ such that $0 \le y \le 1$ and $e_1 y$, $(1-y)e_2$ are small. Then let

$$x_2 = (1 - s_1)^{\frac{1}{2}}z^{\frac{1}{2}}yz^{\frac{1}{2}}(1 - s_1)^{\frac{1}{2}} \text{ and } x_2' = (1 - s_1)^{\frac{1}{2}}z^{\frac{1}{2}}(1 - y)z^{\frac{1}{2}}(1 - s_1)^{\frac{1}{2}}.$$

Remark. Although h is constructed by means of an infinite sum, this sum does not converge in the strict topology of $M(A)$. It seems unusual that we are nevertheless able to prove that h is in $M(A)$.

A^{**} is embedded in $M(A)^{**} \cong A^{**} \oplus (M(A)/A)^{**}$. For a projection p in A^{**} we denote by $\overline{p}^M$ its closure in $M(A)^{**}$ (regarding $M(A)$ as the basic C^*-algebra).

Corollary to Lemma 2. If p and q are closed projections in A^{**} such that $pq = 0$, then $\overline{p}^M \overline{q}^M = 0$.

Sketch of the proof of Theorem 1. Let $p_t = E_{(-\infty,t]}(h)$ and $q_s = E_{[s,\infty)}(k)$.

Then: (1) p_t, q_s are closed
(2) $p_{t_1} \le p_{t_2}$ for $t_1 \le t_2$
(3) $q_{s_1} \ge q_{s_2}$ for $s_1 \le s_2$
(4) $p_t q_s = 0$ for $t < s$.

Let $\tilde{p}_t = \overline{p}_t^M$ and $\tilde{q}_s = \overline{q}_s^M$. Then $\tilde{p}_t, \tilde{q}_s$, which are elements of $M(A)^{**}$, satisfy (1) to (4). One can construct h', k' in $M(A)^{**}$, such that

$E_{(-\infty,t]}(h') = \bigwedge_{t'>t} \tilde{p}_{t'}$ and $E_{[s,\infty)}(k') = \bigwedge_{s'<s} \tilde{q}_{s'}$. Then h',k' have the same properties relative to $M(A)$ as h,k gave relative to A. Since $M(A)$ is unital, q–semicontinuity in $M(A)^{**}$ implies strong semicontinuity in $M(A)^{**}$. Thus Theorem 0 can be applied to find x in $M(A)_{sa}$ such that $h' \geq x \geq k'$. Since the components of h',k' on A^{**} are h,k, this implies that $h \geq x \geq k$.

Remarks.

(1) Although the statement of the main theorem is not as natural as one would like, the theorem seems to be reasonably applicable. Moreover, the apparently most natural conjecture of this type is false ([3]). On the other hand, Lemma 2 is the natural version of Urysohn's lemma for multiplier s, except that perhaps the σ–unitality hypothesis could be weakened (this hypothesis cannot be eliminated altogether).

(2) The proof of Lemma 2 is straightforward enough for us to hope that the technique could have other uses. The rest of the proof is unappealing.

A sketchy description of some of the applications of Theorem 1 follows.

A. In [5] it is proved that if A is σ-unital, then the natural map from $M(A)$ to $M(B)$ is surjective where B is a quotient algebra of A. This and a generalisation in which we use a general (i.e., non-central) closed face of A instead of a quotient algebra, can be deduced easily from Theorem 1.

B. If B is a hereditary C^*-subalgebra of A, let $M(A,B) = B^{**} \cap M(A)$ (the set of multipliers of A which are supported by the open projection of B). Then if B is the smallest hereditary C^*-subalgebra containing B_1 and B_2, where B_1 and B_2 are hereditary and q-commuting, then h is in $M(A,B_1) + M(A,B_2)$ if h is an element of $M(A,B)$ which commutes with at least one of the open projections for B_1 and B_2. An interesting special case is that where B_2 is an ideal and $B_1 + B_2 = A$. Then $M(A) = M(A,B_1) + M(A,B_2)$. This subject was suggested by a question of J. Mingo for the case B_1,B_2 both ideals. G. Pedersen asked us another question on similar lines for the case B_2 an ideal, $B_1 + B_2 = A$. This question was helpful to us and was answered negatively. (See 3.36 of [3] and the remarks following.)

References

1. C.A. Akemann, A Gelfand representation theory for C^*-algebras, *Pac. J. Math.* 39 (1971), 1-11.
2. C.A. Akemann and G.K. Pedersen, Complications of semicontinuity in C^*-algebra theory, *Duke Math. J.* 40 (1973), 785-795.
3. L.G. Brown, Semicontinuity and multipliers of C^*-algebras, preprint.
4. G.K. Pedersen, *C^*-algebras and their automorphism groups*, London, Academic Press, 1979.
5. G.K. Pedersen, SAW*-algebras and corona C^*-algebras, contributions to non-commutative topology, *J. Operator Theory* 15 (1986), 15-32.

Elliptic Invariants and Operator Algebras: Toroidal Examples

R. G. Douglas

The understanding of invariants for elliptic operators on compact closed manifolds is now rather complete based on the index theorem of Atiyah-Singer [1], now almost twenty-five years old, and its refinement and expression in terms of K-homology and cyclic cohomology (cf. [12]). Although the natural extension of linear algebra invariants leads to infinities, one circumvents that by considering the relation of the elliptic operator to the module action of smooth functions. This is successful because the commutator of the elliptic operator and a smooth function is an operator of strictly lower order which can be reduced to a compact operator or even an operator some power of which is trace-class. Invariants for operators which are only partially elliptic or which are defined on non-compact manifolds or on manifolds with boundary are not nearly so well understood although many interesting results have been obtained [10], [9], [19], [3]. Theory, however, is inadequate to cover many examples, even those on tori, which we want to describe in this note. We begin with the classic example for the odd case.

On the torus $\mathbf{T}$, there is only one natural elliptic operator, $D = -i\frac{d}{d\theta}$ on $L^2(\mathbf{T})$. This operator is self-adjoint and hence the natural invariant would be the signature. We let $H_\pm$ denote the positive and negative spectral subspaces for D but encounter problems in trying to define signature since both are infinite dimensional. However, if we let P be the orthogonal projection of $L^2(\mathbf{T})$ onto H_+, then the commutator $[P,\varphi]$ is trace-class for φ in $C^\infty(\mathbf{T})$. Hence, if we let $\mathcal{J}$ be the C^*-algebra generated by the compact operators $\mathcal{K}$ on H_+ and $\{T_\varphi : \varphi \in C(\mathbf{T})\}$, where T_φ is the Toeplitz operator $P\varphi P$ on H_+, then we have the extension

$$0 \longrightarrow \mathcal{K} \longrightarrow \mathcal{T} \longrightarrow C(\mathbf{T}) \longrightarrow 0$$

which defines an element [D] in $\mathcal{K}_1(\mathbf{T})$, the odd K-homology for $\mathbf{T}$ [5]. Or equivalently, we have the element [D] in $KK^1(C(\mathbf{T}),\mathcal{K})$ [18]. In either case, using the K-theory exact sequence arising from the extension or the KK-product, we obtain the index homomorphism $\partial : K_1(C(\mathbf{T})) \longrightarrow K_0(\mathcal{K})$ which yields the formula

$$\text{index } T_\varphi = \partial[\varphi] = \langle [D],[\varphi] \rangle$$

as an element of $K_0(\mathcal{K}) \cong \mathbf{Z}$ (cf. [4]). The classical index theorem for Toeplitz operators which generalizes to the odd Atiyah-Singer index theorem [2] calculates this K-homology element in terms of the homology of $\mathbf{T}$. More recently, Connes has shown [8] how to define the one cyclic cocycle c_D to be a bilinear map on $C^\infty(\mathbf{T})$ such that

$$c_D(\varphi_0,\varphi_1) = \frac{1}{4}\,\text{Tr}\{\varphi_0[F,\varphi_1]\},$$

where Tr denotes the standard trace on $L^2(\mathbf{T})$ and $F = 2P-I$. Then the class $[c_D]$ in $HC^1(C^\infty(\mathbf{T}))$ is the Chern character of [D] in $K^1(C(\mathbf{T})) = K_1(\mathbf{T})$ and the pairing of [D] with $K_0(C^\infty(\mathbf{T})) \cong K_0(C(\mathbf{T}))$ can be expressed:

$$\text{index } T_\varphi = \langle [D],[\varphi] \rangle = c_D(\varphi,\varphi^{-1})$$

for φ in $C^\infty(\mathbf{T})^{-1}$. The formula can be extended to φ in $M_k(C^\infty(\mathbf{T}))$ by substituting $\text{Tr}\cdot\text{Tr}_k$ for Tr, where Tr_k denotes the normalized trace on $M_k(\mathbb{C})$.

Whether expressed in odd K-homology, odd KK-theory or odd cyclic cohomology, we have identified the basic primary invariant for D [12].

Let us now consider the two torus T^2. Here there is a family of natural operators but only two natural elliptic operators ∂ and $\bar{\partial}$, at least up to changes in the metric on T^2. Actually, more natural from many points of vie is the elliptic operator

$$D_{T^2} = \begin{pmatrix} 0 & \partial \\ \bar{\partial} & 0 \end{pmatrix}$$

which is the self-adjoint Dirac operator defined on $L^2(T) \oplus L^2(T)$. The operator $\bar{\partial}$ can be used to define $[\bar{\partial}]$ in $K_0(T^2) = K^0(C(T^2))$ [12], the even K-homology of T^2, or equivalently, D_{T^2} can be used to define $[D_{T^2}]$ in $KK^0(C(T^2),\mathcal{K}) \cong K^0(C(T^2))$ [17]. The pairing of $K_0(C(T^2))$ with $K^0(C(T^2))$ expresses the index of the operator $\bar{\partial}_E$ defined by $\bar{\partial}$ with coefficients in the complex vector bundle E over T^2. Hence, we have

$$\text{index } \bar{\partial}_E = \langle[\bar{\partial}],[E]\rangle = [E]\cdot[\bar{\partial}],$$

where [E] denotes the class in $K^0(T^2) = K_0(C(T^2)) = KK^0(\mathbb{C},C(T^2))$ [18].

Again, we can use D_{T^2} to define a two cyclic cocycle $c_{D_{T^2}}$ on $C^\infty(T^2)$ whose class in $HC^2(C^\infty(T^2))$ is its Chern character [8]. We omit technicalities but basically $c_{D_{T^2}}$ is a linearization of a standard formula expressing index necessitated by the fact that the commutators are not trace-class but only Hilbert-Schmidt. Thus following Connes [8] we have

$$c_{D_{T^2}}(\varphi_0,\varphi_1,\varphi_2) = \frac{1}{8}\,\mathrm{Tr}_s\{\tilde{\varphi}_0[F_{T^2},\tilde{\varphi}_1][F_{T^2},\tilde{\varphi}_2]\}$$

with $\tilde{\varphi}_i = \begin{pmatrix} \varphi_i & 0 \\ 0 & \varphi_i \end{pmatrix}$ defined on $L^2(T^2) \oplus L^2(T^2)$ for φ_i in $C^\infty(T^2)$, F_{T^2} the symmetry in the polar form for D_{T^2}, and Tr_s the super trace defined on $\mathcal{L}(L^2(T^2) \oplus L^2(T^2))$ such that

$$\mathrm{Tr}_s \begin{pmatrix} A & B \\ C & D \end{pmatrix} = \mathrm{Tr}\,A - \mathrm{Tr}\,D,$$

where A,B,C and D are trace class operators on $L^2(T^2)$. We extend $c_{D_{T^2}}$ to $M_k(C^\infty(T^2))$ using the normalized trace on $M_k(\mathbb{C})$. For φ a projection-valued $M_k(C^\infty(T^2))$, range φ is a complex vector bundle E_φ on T^2, and we set

$$\text{index}\ \bar{\partial}_{E_\varphi} = [E_\varphi]\cdot[D_{T^2}] = c_{D_{T^2}}(\varphi,\varphi,\varphi).$$

Again, whether $[\bar{\partial}]$, $[D_{T^2}]$, or $[c_{D_{T^2}}]$, we have expressed the basic even primary invariant for D_{T^2} [12].

We have described the odd and even invariants for D and D_{T^2}, respectively. Although we could try to define the opposite invariants by considering the maps

$$K_0(C(T)) \longrightarrow K_1(\mathcal{K})$$

$$K_1(C(T^2)) \longrightarrow K_1(\mathcal{K})$$

defined using the KK-product, this would come to nought since $K_1(\mathcal{K}) = 0$. Seeking cyclic cohomology classes would not accomplish anything either, since both the even class for D and the odd class for D_{T^2} are zero. We will see for non-elliptic operators, that there normally is both an even and an odd class.

We now want to consider other operators on T^2. The simplest are the $D_j = -i\frac{\partial}{\partial\theta_j}$ for j = 1,2, which are both self-adjoint on $L^2(T^2)$ but neither is elliptic. Thus for φ in $C(T^2)$ the commutators $[F_1,\varphi]$ and $[F_2,\varphi]$ are not compact, where F_j is the symmetry in the polar form for D_j, j = 1,2, but can be seen to be in $\mathcal{K}(L^2(T)) \otimes C(T) = \mathcal{C}_1$ and $C(T) \otimes \mathcal{K}(L^2(T)) = \mathcal{C}_2$, respectively acting on $L^2(T^2)$. If we let $\mathcal{T}_j$ denote the C^*-algebra generated by $\mathcal{C}_j$ and $\{P_j\varphi P_j : \varphi \in C(T^2)\}$, where $P_j = \frac{1}{2}(F_j + I)$, then we have the short exact sequences

$$0 \longrightarrow \mathcal{C}_j \longrightarrow \mathcal{T}_j \longrightarrow C(T^2) \longrightarrow 0$$

and hence the element $[D_j]$ in $KK^1(C(T^2), \mathcal{C}_j)$ for j = 1,2. Now $[D_j]$ can be used to define the index map

$$K_1(C(T^2)) \longrightarrow K_0(\mathcal{C}_j)$$

which expresses the index of the Toeplitz operator $P_j\varphi P_j$. The latter agrees with the ordinary index of the ordinary Toeplitz operator T_{χ_j}, where χ_j is φ restricted to $T \times e^{i\theta}$ for j = 1 or $e^{i\theta} \times T$ for j = 2. This follows because

$$K_0(\mathcal{C}_j) \cong K_0(\mathcal{K}) \cong \mathbb{Z}.$$

This index can also be expressed in terms of the odd cyclic cocycle c_{D_1} defined by

$$c_{D_1}(\varphi_0, \varphi_1) = \frac{1}{8\pi}\int_0^{2\pi} \mathrm{Tr}\{\varphi_0(.,e^{i\theta})[F_1, \varphi_1(.,e^{i\theta})]\}d\theta$$

and the odd cyclic cocycle c_{D_2} defined by the same formula with the two variables switched. The justification for the "trace" in this formula will become clearer.

Since $K_1(\mathcal{C}_j) \cong \mathbb{Z}$, this is a case where there is the second index map

$$K_0(C(T^2)) \longrightarrow K_1(\mathcal{C}_j) \cong \mathbb{Z}.$$

The meaning of this map can be seen by taking a bundle E over T^2, restricting it to a circle fiber and considering the self-adjoint operator defined by D_j with coefficients in the vector bundle obtained for each point of the other circle. This circle of self-adjoint operators defines an element of $K^1(T)$ which is the index of $[E]\cdot[D]$ in $K_1(\mathcal{C}_2) \cong K_1(C(T)) = K^1(T)$. Thus, in this case the operators D_j have two index maps from the K-theory of $C(T^2)$ to the K-theory of $\mathcal{C}_j$.

These index maps result because D_1 is longitudinally elliptic along the curves $T \times e^{i\theta}$ of the product foliation $\mathcal{F}_1$ of T^2 and D_2 is along the leaves $e^{i\theta} \times T$ of $\mathcal{F}_2$. The commutators $[F_j, \varphi]$ lie in the foliation C^*-algebra $C^*(\mathcal{F}_j)$ of Connes (cf. [7]). In general, the foliation algebra consists of those operators which are "compact-like" along the leaves and a continuous multiplier in the transverse direction. In this case, $C^*(\mathcal{F}_j) = \mathcal{C}_j$ and the smooth foliation algebra $C^\infty(\mathcal{F}_j)$ is contained in $C^\infty(T, \mathcal{L}^1)$, where $\mathcal{L}^1$ denotes the algebra of trace-class operators. The space $\mathbf{T}$ is a transversal and Haar measure on it induces a trace on the foliation algebra which coincides with that used in defining the odd cyclic cocycle c_{D_j} above.

There is another way to obtain invariants for the operators D_j and that is by using the fact that they are transversally elliptic relative to the opposite foliations, that is, D_1 for $\mathcal{F}_2$ and D_2 for $\mathcal{F}_1$. We can also provide more concrete motivation as follows. For φ in $C^\infty(\mathbf{T}^2)$ the commutator $[F_1,\varphi]$ is compact-like along the fibre $\mathbf{T} \times e^{i\theta}$ but not in the transverse direction $e^{i\theta} \times \mathbf{T}$. However, if instead of using φ, we used k in $C^*(\mathcal{F}_2)$ which is already "compact-like" in the direction $e^{i\theta} \times \mathbf{T}$, we obtain a commutator $[F_1,k]$ which is in $\mathcal{K}(L^2(\mathbf{T}^2))$. Therefore, if we let $\mathcal{J}_1^\sharp$ denote the C^*-algebra generated by $\mathcal{K}(L^2(\mathbf{T}^2))$ and $\{P_1kP_1 : k \in C^*(\mathcal{F}_2)\}$, then we obtain the short exact sequence

$$0 \longrightarrow \mathcal{K} \longrightarrow \mathcal{J}_1^\sharp \longrightarrow C^*(\mathcal{F}_2) \longrightarrow 0$$

and hence we obtain $[D_j]^\sharp$ in $KK^1(C^*(\mathcal{F}_2),\mathcal{K})$ which can be used to define an index map

$$K_1(C^*(\mathcal{F}_2)) \longrightarrow K_0(\mathcal{K}) \cong \mathbb{Z}.$$

Similarly, we have $[D_j]^\sharp$ in $KK^2(C^*(\mathcal{F}_1),\mathcal{K})$ and the index map

$$K_1(C^*(\mathcal{F}_1)) \longrightarrow K_0(\mathcal{K}) \cong \mathbb{Z}.$$

Although these index maps yield the ordinary index for the "Toeplitz operators" P_jkP_j defined for the multiplier k in the transverse foliation algebra, we have introduced them for a different purpose. First, we need to point out that odd transverse cyclic cocycles can be defined using the same formula as for the longitudinal cyclic cocycles - only the domain and the trace have changed. Namely we define

$$c^{\not{A}}_{D_j}(k_0,k_1) = \frac{1}{4}\mathrm{Tr}\{k_0[F_j,k_1]\}$$

for k_0 and k_1 in $C^\infty(\mathcal{F}_{j'})$, where $j' = \begin{cases} 1 & j = 2 \\ 2 & j = 1 \end{cases}$. Then $[c^{\not{A}}_{D_j}]$ lies in $HC^1(C^\infty(\mathcal{F}_{j'}))$ and defines the index pairing of $[D_j]^{\not{A}}$ and $K_1(C^*(\mathcal{F}_{j'}))$. This is the odd analogue of the transverse index theory of Connes [9].

Now we can form the sharp product of D_1 and D_2 to obtain the differential operator $D_1 \# D_2$ which can be identified with D_{T^2}. We are interested in factoring the index invariants for D_{T^2} in terms of those for D_1 and D_2. Since even cocycles pair with even K-theory and odd with odd, we can form the diagram

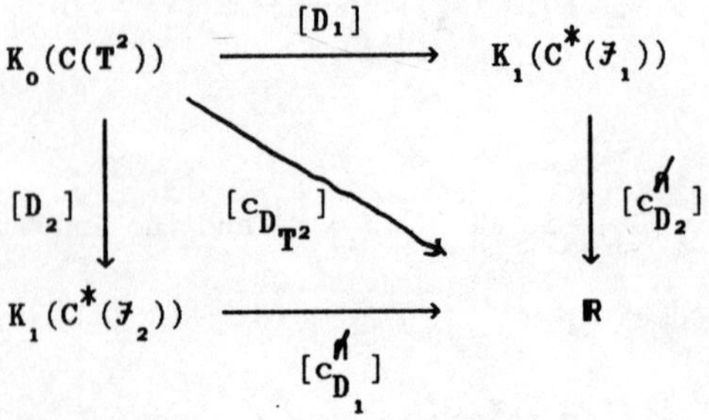

$$\begin{array}{ccc} K_0(C(T^2)) & \xrightarrow{[D_1]} & K_1(C^*(\mathcal{F}_1)) \\ {\scriptstyle [D_2]}\downarrow & \searrow {\scriptstyle [c_{D_{T^2}}]} & \downarrow {\scriptstyle [c^{\not{A}}_{D_2}]} \\ K_1(C^*(\mathcal{F}_2)) & \xrightarrow[{[c^{\not{A}}_{D_1}]}]{} & \mathbb{R} \end{array}$$

which can be shown to commute using the associativity of the KK-product and the naturality properties of the cyclic cohomology-valued Chern character. This is rather straight-forward but we seek a "Fubini Theorem" for the cyclic cocycles involved. That is, the trace, a zero cyclic cocycle on $\mathcal{L}^1(L^2(T^2))$, induces the two cyclic cocycle $c_{D_{T^2}}$ on $C^\infty(T^2)$. We would like to obtain it in two steps through $C^\infty(\mathcal{F}_1)$ and through $C^\infty(\mathcal{F}_2)$. Strictly speaking this would involve a "bivariant cyclic theory" but in this case we can more or less avoid it. We proceed as follows.

The trace on $\mathcal{L}^1(L^2(T^2))$ induces the one cyclic cocycle $c_{D_j}^{\not{A}}$ on $C^\infty(\mathcal{F}_{j'})$. Using a fixed rank one projection in $\mathcal{L}^1$, we can define a *-homomorphism

$$C^\infty(T) \longrightarrow C^\infty(\mathcal{F}_{j'})$$

which enables us to pull $c_{D_j}^{\not{A}}$ back to a one cyclic cocycle on $C^\infty(T)$. Using the one cyclic cocycle c_D also defined on $C^\infty(T)$ by $-i\frac{d}{d\theta}$, we can use Conne's product [9] to obtain a two cyclic cocycle on the tensor product $C^\infty(T) \otimes C^\infty(T)$ which agrees with $c_{D_{T^2}}$. Thus we have a recipe for constructing the two cyclic cocycle on $C^\infty(T^2)$ as an iteration of one cyclic cocycles. But we had to use the identification of $C^\infty(\mathcal{F}_{j'})$ with $C^\infty(T) \otimes \mathcal{L}^1$ which is possible in this case because the space of leaves for $\mathcal{F}_{j'}$ is the manifold T. For that reason we need a direct way of using the one cyclic cocycle on $C^\infty(\mathcal{F}_{j'})$ plus the longitudinal operator $D_{j'}$ to construct $c_{D_{T^2}}$. Perhaps one should seek to construct the dual map in cyclic homology $HC_2(C^\infty(T^2)) \longrightarrow HC_1(C^\infty(\mathcal{F}_{j'}))$ generated by the "slash product" with $c_{D_{j'}}$ in $HC^1(C^\infty(\mathcal{F}_j))$. Possibly one can identify the tensor product $C^\infty(T^2) \otimes_{C(T^2)} C^\infty(\mathcal{F}_2)$ with $C^\infty(T^2) \otimes \mathcal{L}^1(L^2(T^2))$ in a canonial way and extend the cyclic cocycle product to this setting.

The need for understanding iterations such as that described above is made clear by considering invariants for the family of self-adjoint operators $D_\alpha = -i(\frac{\partial}{\partial\theta_1} + \alpha\frac{\partial}{\partial\theta_2})$ on T^2. If $\mathcal{F}_\alpha$ denotes the foliation of T^2 with leaves the lines of slope α, then $C^*(\mathcal{F}_\alpha)$ is isomorphic to $C(T) \otimes \mathcal{K}(L^2(T))$ for α rational but is not a type I algebra for α irrational. Again Haar measure on the transversal T induces a trace Tr_α on $C^\infty(\mathcal{F}_\alpha)$ and we can define both longitudinal and transverse cyclic cocycles for D_α. Since D_α is longitudinally elliptic for $\mathcal{F}_\alpha$, the commutator* $[F_\alpha,\varphi]$ is in $C^*(\mathcal{F}_\alpha)$ and if we let $\mathcal{T}_\alpha$ be the C^*-algebra generated by $C^*(\mathcal{F}_\alpha)$ and $\{P_\alpha\varphi P_\alpha : \varphi \in C(T^2)\}$, where F_α is the symmetry in the polar form for D_α and $P_\alpha = \frac{1}{2}(I + F_\alpha)$, then we obtain the short exact sequence

$$0 \longrightarrow C^*(\mathcal{F}_\alpha) \longrightarrow \mathcal{T}_\alpha \longrightarrow C(T^2) \longrightarrow 0$$

yielding an element $[D_\alpha]$ in $KK^1(C(T^2),C^*(\mathcal{F}_\alpha))$. The resulting index formula

$$K_1(C(T^2)) \longrightarrow K_0(C^*(\mathcal{F}_\alpha))$$

yields the index for the Toeplitz operators $T_\varphi = P_\alpha\varphi P_\alpha$. Pairing with the trace Tr_α on $C^\infty(\mathcal{F}_\alpha)$ yields a real-valued index that agrees with that introduced in [6]. The trace also enables us to define a longitudinal one cyclic cocycle c_{D_α} on $C^\infty(T^2)$ by

---

*Strictly speaking this is not true and one must introduce "smooth projections" but we do not go into that (cf. [13], [14])

$$c_{D_\alpha}(\varphi_0,\varphi_1) = \frac{1}{4}\mathrm{Tr}_\alpha\{\varphi_0[F_\alpha,\varphi_1]\}.$$

Again, we have $c_{D_\alpha}(\varphi,\varphi^{-1})$ equal to the real-valued index of T_φ.

Since the operator D_α is transversally elliptic to $\mathcal{F}_\beta$ for $\beta \neq \alpha$, we have a transverse index in these cases. This follows since the commutator* $[F_\alpha,k]$ will be a compact operator on $L^2(T^2)$ for k in $C^*(\mathcal{F}_\beta)$ and indeed will be trace class for k in $C^\infty(\mathcal{F}_\beta)$. Thus we can define the transverse K-homology class $[D_\alpha]^{\pitchfork}$ in $K^1(C^*(\mathcal{F}_\beta))$ from the short exact sequence

$$0 \longrightarrow \mathcal{K} \longrightarrow \mathcal{J}_{\alpha,\beta} \longrightarrow C^*(\mathcal{F}_\beta) \longrightarrow 0,$$

where $\mathcal{J}_{\alpha,\beta}$ is the C^*-algebra generated by $\mathcal{K}$ and $\{P_\alpha k P_\alpha : k \in C^\alpha(\mathcal{F}_\beta)\}$. This defines an index map $K_1(C^\alpha(\mathcal{F}_\beta)) \longrightarrow K_0(\mathcal{K}) \cong \mathbb{Z}$ for which there also exists the one cyclic cocycle $c^{\pitchfork}_{D_\alpha}$ on $C^\infty(\mathcal{F}_\beta)$ defined by

$$c^{\pitchfork}_{D_\alpha}(k_0,k_1) = \frac{1}{4}\mathrm{Tr}\{k_0[F_\alpha,k_1]\}$$

for k_0 and k_1 in $C^\infty(\mathcal{F}_\beta)$.

Although the geometry and the algebras are more complicated for the general foliations, we still have the following index diagram

---

*See the previous footnote.

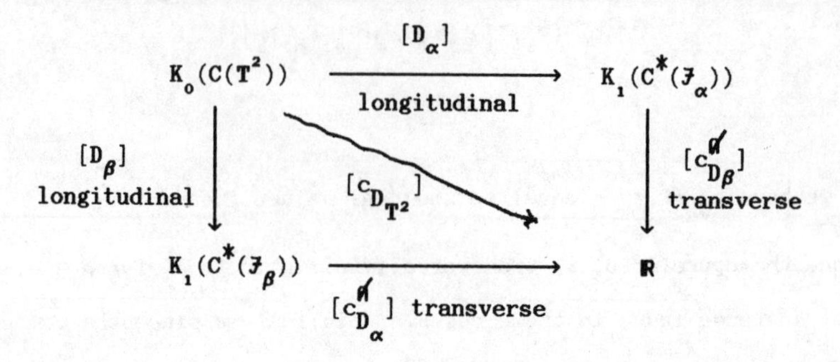

which expresses the index map for D_{T^2} as an iterated map with respect to the two foliations $\mathcal{F}_\alpha$ and $\mathcal{F}_\beta$. (Care must be taken to make certain that the metrics match up.) The same discussion as above applies to the problem of iterating the two cyclic cocycle $c_{D_{T^2}}$ in terms of D_α and D_β except that when α and β are irrational, we can not replace $C^\infty(\mathcal{F}_\alpha)$ by the algebra for the space of leaves. Therefore, something more is necessary to proceed further in this case.

We leave this now to discuss a case which is not fully elliptic. Namely we seek to relate $c_{D\alpha}$ and $c^{\mathcal{F}}_{D\alpha}$, where the latter cyclic cocycle is defined relative to the product foliation $\mathcal{F}_2$. Recall that the formulas are the same, only the domain and trace are different. There are reasons for wanting to relate them and this is the heart of my work with Hurder and Kaminker [13], [15]. We seek to "pull back" $c^{\mathcal{F}}_{D\alpha}$ from $C^\infty(\mathcal{F}_2)$ to $C^\infty(T^2)$. Since there is no homomorphism from $C^\infty(T^2)$ to $C^\infty(\mathcal{F}_2)$, we use an asymptotic one. Let $\Delta = -\frac{\partial^2}{\partial\theta_2^2}$ denote the Laplacian for $\mathcal{F}_2$ and let

$$\psi_t : C^\infty(T^2) \longrightarrow C^\infty(\mathcal{F}_2)$$

be defined by $\psi_t(\varphi) = \dfrac{e^{-t\Delta}\varphi}{\operatorname{Tr} e^{-t\Delta}}$, where Tr is the ordinary trace calculated on a fiber. One can show that ψ_t is well-defined and that it is asymptotically multiplicative, that is, that $\lim_{t\to 0^+} \|\psi_t(\varphi_1\cdot\varphi_2) - \psi_t(\varphi_1)\psi_t(\varphi_2)\| = 0$. Moreover, one can show that

$$(Tc^{\not{A}}_{D\alpha})(\varphi_0,\varphi_1) = \lim_{t\to 0^+} c^{\not{A}}_{D\alpha}(\psi_t(\varphi_0),\psi_t(\varphi_1))$$

exists and**

$$Tc^{\not{A}}_{D\alpha} = c_{D\alpha}.$$

One can also show that

$$\psi_t : C^\infty(\mathcal{F}_\alpha) \longrightarrow \mathcal{L}^1,$$

that

$$(T\ \operatorname{Tr})(k) = \lim_{t\to 0^+} \operatorname{Tr}(\psi_t(k))$$

exists and

$$(T\ \operatorname{Tr})(k) = \operatorname{Tr}_\alpha k.$$

Therefore we have the Fubini square

**Actually a double limit is required to remove the approximation to the projection.

$$\begin{array}{ccc} c_{D_\alpha} \in HC^1(C^\infty(T^2)) & \xleftarrow[\text{longitudinal}]{D_\alpha} & \mathrm{Tr}_\alpha \in HC^0(C^\infty(\mathcal{F}_\alpha)) \\ \uparrow \text{longitudinal renormalization} & & \uparrow \text{transverse renormalization} \\ c'_{D_\alpha} \in HC^1(C^\infty(\mathcal{F}_2)) & \xleftarrow[\text{transverse}]{D_\alpha} & \mathrm{Tr} \in HC^0(\mathcal{L}^1) \end{array}$$

This is the diagram which results in a non fully elliptic situation in which renormalization or some sort of averaging is necessary.

We will discuss briefly two other cases which illustrate new phenomenon. The first is on T^3 for the operator

$$D_{\underline{\alpha}} = i(\alpha_1 \frac{\partial}{\partial\theta_1} + \alpha_2 \frac{\partial}{\partial\theta_2} + \alpha_3 \frac{\partial}{\partial\theta_3})$$

with related foliation $\mathcal{F}_{\underline{\alpha}}$ of T^3. Assume that α_1, α_2 and α_3 are linearly independent over $\mathbb{Z}$. Again, since $D_{\underline{\alpha}}$ is longitudinally elliptic, the commutator $[F_{\underline{\alpha}}, \varphi]$ is in* $C^*(\mathcal{F}_{\underline{\alpha}})$ where $F_{\underline{\alpha}}$ is the symmetry in the polar form for $D_{\underline{\alpha}}$ and φ is in $C(T^3)$. And, if $\mathcal{J}_{\underline{\alpha}}$ is the C^*-algebra generated by $C^*(\mathcal{F}_{\underline{\alpha}})$ and $\{P_{\underline{\alpha}}\varphi P_{\underline{\alpha}} : \varphi \in C(T^3)\}$, then we obtain the short exact sequence

$$0 \longrightarrow C^*(\mathcal{F}_{\underline{\alpha}}) \longrightarrow \mathcal{J}_{\underline{\alpha}} \longrightarrow C(T^3) \longrightarrow 0$$

---

*See the first footnote.

yielding the element $[D_{\underline{\alpha}}]$ in $KK(C(T^3), C^*(\mathcal{F}_{\underline{\alpha}}))$. Since $D_{\underline{\alpha}}$ is the Dirac operator for $\mathcal{F}_{\underline{\alpha}}$, the resulting index map

$$K_1(C(T^3)) \longrightarrow K_0(C^*(\mathcal{F}_{\underline{\alpha}}))$$

is an isomorphism [16], [14]. However, the real-valued index obtained from the trace $Tr_{\underline{\alpha}}$ on $C^*(\mathcal{F}_{\underline{\alpha}})$ is not an isomorphism. In fact, one can show that the real-valued index depends only on $H^1(T^3,\mathbb{Z}) = \mathbb{Z} \oplus \mathbb{Z} \oplus \mathbb{Z}$, whereas the Chern character on $K_1(C(T^3))$ has an $H^3(T^3,\mathbb{Z}) = \mathbb{Z}$ summand. This can be viewed as a reformulation of the result of Schaeffer [20] that the real-valued index for almost periodic Wiener-Hopf operators depends only on the determinant of the mean-motion.

However, from our vantage, we can see how to detect the entire index class for these operators. The transversal to $\mathcal{F}_{\underline{\alpha}}$ is now T^2 and Haar measure isn't the only invariant even cyclic-cohomology on $C^\infty(T^2)$. In particular, the Dirac operator defines $c_{D_{T^2}}$ which we discussed earlier. The trick is to use $c_{D_{T^2}}$ to define a two cyclic-cocycle on the foliation algebra $C^\infty(\mathcal{F}_{\underline{\alpha}})$ in analogy with what is done for zero cyclic-cocycles or traces. Although it is not obvious how to do this in general, here we can use the differential operator D_{T^2}. If we extend this elliptic operator on T^2 to be $I \otimes D_{T^2}$ on T^3 (which we will denote D_{T^2}), then we obtain a transversely elliptic operator for $\mathcal{F}_{\underline{\alpha}}$. The two cyclic-cocycle $c^{\not\alpha}_{D_{T^2}}$ is the one we seek and the index map

$$K_1(C(T^3)) \xrightarrow{[D_{\underline{\alpha}}]} K_0(C^*(\mathcal{F}_{\underline{\alpha}})) \xrightarrow{[c^{\not\alpha}_{D_{T^2}}]} \mathbb{Z}$$

detects the $H^3(T^3,\mathbb{Z})$ summand.

As I suspect the reader has already noticed, this composition is part of a fully-elliptic square

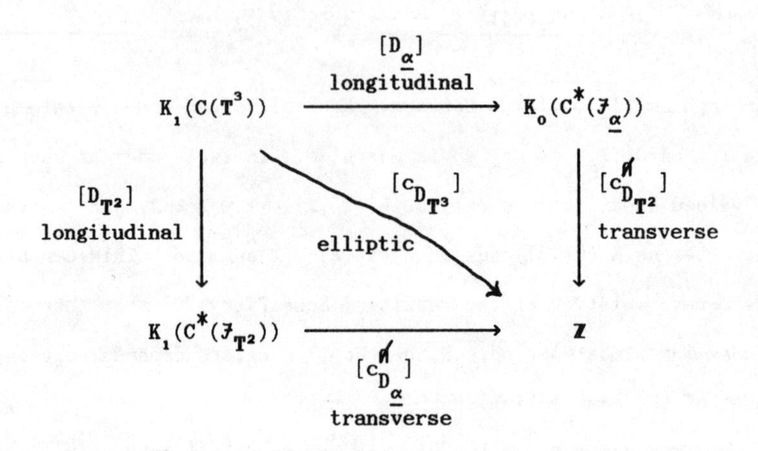

where $\mathcal{F}_{T^2}$ denotes the product foliation $T \times T^2$ with leaves $e^{i\theta} \times T^2$ and D_{T^3} is the Dirac operator on T^3 corresponding to the sharp product $D_{\underline{\alpha}} \# D_{T^2}$.

Finally, suppose we move the previous example up one more dimension to the case of the operator $D_{\underline{\alpha}} = i(\alpha_1\frac{\partial}{\partial\theta_1} + \alpha_2\frac{\partial}{\partial\theta_2} + \alpha_3\frac{\partial}{\partial\theta_3} + \alpha_4\frac{\partial}{\partial\theta_4})$ on T^4 with related foliation $\mathcal{F}_{\underline{\alpha}}$, where the α's are assumed to be linearly independent over $\mathbb{Z}$. Without going into any detail we can repeat the previous discussion except for a major problem on either path from $K_1(C(T^4))$ to $\mathbb{R}$. Since $D_{\underline{\alpha}}$ and D_{T^2} are not a fully elliptic pair, renormalization is required. One way to accomplish this is to consider an intermediate foliation. For example, the transverse one cyclic cocycle $c^{\#}_{D_{\underline{\alpha}}}$ on $C^\infty(\mathcal{F}_{T^3})$ can be renormalized using the Laplace operator on T to the one cyclic cocycle $Tc^{\#}_{D_{\underline{\alpha}}}$ on $C^\infty(\mathcal{F}_{T^2})$ which then pairs with the index map $K_1(C(T^4)) \longrightarrow K_1(C^*(\mathcal{F}_{T^2}))$ defined by $[D_{T^2}]$. Similarly, we can either renormalize the ordinary trace on $\mathcal{L}^1(L^2(T^4))$ using the Laplace operator on the same factor and then define the transverse two-cyclic cocycle on $C^\infty(\mathcal{F}_{\underline{\alpha}})$ or one could try to proceed in the other order. In any case it is obvious that much of this appears ad hoc and needs further study.

Finally, before concluding let us point out that $H^3(T^4,\mathbb{Z}) = \mathbb{Z} \oplus \mathbb{Z} \oplus \mathbb{Z}$ which corresponds to the three different choices of two circles from three in the fiber T^3 in the product foliation $\mathcal{F}_{T_3}$. Moreover, the two cyclic cocycles defined by each choice corresponds to the different factor of $\mathbb{Z}$ in $H^3(T^4,\mathbb{Z})$ and all are needed to fully detect the index map

$$K_1(C(T^4)) \longrightarrow K_0(C^*(\mathcal{F}_{\underline{\alpha}})).$$

As one can see things get more complicated with increasing transverse dimension but I hope that these toroidal examples have indicated some of the possible structure.

In my work with Hurder and Kaminker [13], [15], we have shown how secondary invariants enter for renormalized cyclic cocycles. One would expect higher terms in the eta invariant series to appear in the above examples or at least for the analogue when the torus is replaced by more complicated spaces.

State University of New York at Stony Brook

REFERENCES

[1] M. F. Atiyah and I. M. Singer, The index of elliptic operators I, Ann. Math. 87 (1968) 484-530.

[2] P. Baum and R. G. Douglas, Toeplitz operators and Poincaré duality. In Proc. Toeplitz Memorial Conference, Tel Aviv, 1981 (ed. I. C. Gohberg) Basel, Birkhauser (1982) 137-66.

[3] P. Baum, R. G. Douglas, and M. J. Taylor, Cycles and relative cycles in analytical K-homology, J. Diff. Geom., to appear.

[4] B. Blackadar, K-Theory for Operator Algebras, Math Sci. Res. Inst. Publ. No. 5, Springer Verlag, New York (1986).

[5] L. G. Brown, R. G. Douglas, and P. A. Fillmore, Extensions of C^*-algebras and K-homology, Ann. Math. (2) 105 (1977) 265-324.

[6] L. A. Coburn, R. G. Douglas, D. G. Schaeffer, and I. M. Singer, On C^*-algebras of operators on a half-space II. Index theory, Inst. Hautes Etudes Sci. Pub. Math. 40 (1971) 69-79.

[7] A. Connes, A survey of foliations and operator algebras, In Operator Algebras and Applications (ed. R. V. Kadison), Proc. Symp. Pure Math. 38 Amer. Math. Soc., Providence (1981).

[8] A. Connes, Non-commutative differential geometry, Publ. Inst. Hau. Etu. Sci. 62 (1985) 41-144.

[9] A. Connes, Cyclic cohomology and the transverse fundamental class of a foliation, In Geometric methods in operator algebras (ed. H. Araki and E. G. Effros), Pitman Research Notes 123 Longman, Harlow (1986).

[10] A. Connes and G. Skandalis, The longitudinal index theorem for foliations, Publ. Res. Inst. Math. Sci. Kyoto Univ. 20 (1984), no. 6, 1139-1183.

[11] R. G. Douglas, C^*-algebra extensions and K-homology, Ann. Math. Studies 95, Princeton (1980).

[12] ——————, Invariant theory for elliptic operators, Proc. R. Ir. Acad. 86A (1986) 161-174.

[13] R. G. Douglas, S. Hurder, and J. Kaminker, Toeplitz operators and the eta invariant: the case of S^1, In Index Theory of Elliptic Operators, Foliations, and Operator Algebras, (ed. C. Schochet), Contemp. Math. 70 (1988).

[14] ——————, The longitudinal cyclic cocycle and the index of Toeplitz operators, in preparation.

[15] ———————, Cyclic cocycles, renormalization and von Neumann eta invariants, in preparation.

[16] R. Ji and J. Kaminker, The K-theory of Toeplitz extensions, J. Operator Thy., to appear.

[17] R. Ji and J. Xia, On the classification of commutator ideals, J. Funct. Anal., to appear.

[18] G. G. Kasparov, The operator K-functor and extensions of C^*-algebras, Izu. Akad. Nauk SSSR, Ser. Math. 44 (1980), 571-636; Math USSR Izvestiya 16 (1981) 513-572.

[19] J. Roe, An index theorem for open manifolds I, J. Diff. Geom., to appear.

[20] D. Schaeffer, An index theorem for systems of difference operators on a half space, Publ. Math. I.H.E.S., 1973-397-403.

On Multilinear Double Commutant Theorems

Edward G. Effros
Ruy Exel

1. Introduction

Let us suppose that $\mathcal{B}=\mathcal{B}(H)$ is the von Neumann algebra of bounded operators on a Hilbert space H. It follows from the GNS construction that if ω is a normal linear functional on $\mathcal{B}$ with $\|\omega\|\leq 1$, then there exist vectors $\xi,\eta\in H^{\infty}=H\oplus H\oplus\ldots$ for which $\|\xi\|,\|\eta\|\leq 1$ and

$$\omega(r) = (r\otimes I)\eta\cdot\xi \quad (r\in\mathcal{B}),$$

where $r\mapsto r\otimes I$ is the diagonal representation of $\mathcal{B}$ on H^{∞}. We shall refer to this as a GNS representation for ω. Although the vectors ξ and η are not themselves unique, they determine a unique element $\xi^{*}\odot\eta = \sum\xi_j^{*}\otimes\eta_j$ in the projective tensor product $H^{*}\hat{\otimes}H$. Letting $\mathcal{B}_{*}$ denote the predual of $\mathcal{B}$, the GNS representation determines in this manner the well-known isometry

$$\mathcal{B}_{*}\cong H^{*}\hat{\otimes}H.$$

A fact that has been overlooked, or at least underemphasized, is that if $\mathcal{R}\subseteq\mathcal{B}$ is a von Neumann algebra with predual $\mathcal{R}_{*}$ and commutant $\mathcal{R}'$, then we have a corresponding identification

$$\mathcal{R}_{*}\cong H^{*}\hat{\otimes}_{\mathcal{R}'}H. \tag{1.1}$$

The tensor product on the right is obtained by regarding H (resp., H^{*}) as a left (resp., right) $\mathcal{R}'$ module, and then forming the obvious Banach $\mathcal{R}'$-module projective tensor product (see below). We may regard (1.1) as a description of the non-uniqueness of the GNS representation for normal functionals on $\mathcal{R}$. An important aspect of (1.1) is that it is equivalent to the von Neumann Double Commutant Theorem.

Recently Christensen and Sinclair have discovered a remarkable multivariable generalization of the GNS representation [2]. Given a completely contractive multilinear function $F:\mathcal{R}\times\ldots\times\mathcal{R}\to\mathcal{B}(K)$ which is normal in each variable, there exist index sets $J_1,\ldots,J_n$ and a diagram of "bridging maps"

$$(1.2)\qquad K \underset{S_n}{\longrightarrow} H^{J_n} \underset{S_{n-1}}{\longrightarrow} \ldots \longrightarrow H^{J_1} \underset{S_0}{\longrightarrow} K$$

which are contractive and satisfy

$$(1.3)\qquad F(r_1,\ldots,r_n) = S_0(r_1\otimes I)S_1\ldots S_{n-1}(r_n\otimes I)S_n.$$

We also leWe are particularly interested in the case that $K=\mathbb{C}$, i.e., the scalar valued functions F. For such functions we have the simplified expression

$$(1.4)\qquad F(r_1,\ldots,r_n) = (r_1\otimes I)S_1\ldots S_{n-1}(r_n\otimes I)\eta\cdot\xi,$$

where we may assume that $\eta,\xi\in H^\infty=H^{J_1}=H^{J_n}$.

Given von Neumann algebras $\mathcal{R}$ and $\mathcal{S}$, we denote the space of all completely bounded multilinear maps $F:\mathcal{R}\times\ldots\times\mathcal{R}\to\mathcal{S}$ by $\mathcal{M}(\mathcal{R}^n,\mathcal{S})$, and those which are normal in each variable by $\mathcal{M}^\sigma(\mathcal{R}^n,\mathcal{S})$. We let $\mathcal{M}(\mathcal{R})=\mathcal{M}(\mathcal{R},\mathcal{R})$ and $\mathcal{M}^\sigma(\mathcal{R})=\mathcal{M}^\sigma(\mathcal{R},\mathcal{R})$.

In this paper we investigate the non-uniqueness of the representation (1.4). Our goal is to find an analogue of (1.1) for the Banach space

$$\mathcal{R}^n{}_* = \mathcal{M}^\sigma(\mathcal{R}^n,\mathbb{C}).$$

The dual of this space is the <u>normal Haagerup tensor product</u>

$$\mathcal{R}\otimes_h{}^\sigma\mathcal{R}\otimes_h{}^\sigma\ldots\otimes_h{}^\sigma\mathcal{R},$$

which is of interest in the theory of completely bounded normal cohomology (see [5] and [1]). In §3 we prove that

$$(1.5)\qquad \mathcal{R}^2{}_* \cong H^*\bar{\otimes}_{\mathcal{R}'}\mathcal{B}\bar{\otimes}_{\mathcal{R}'}H,$$

where $\bar{\otimes}_{\mathcal{R}'}$ denotes a natural Banach module tensor product. This result leads to a much simpler proof of the "Bilinear Double Commutant Theorem", which was first considered in [5]. Letting $\mathcal{B}=\mathcal{B}(H)$ and $\mathcal{M}_{\mathcal{R}'}(\mathcal{B})$ be the completely bounded $\mathcal{R}'$ bimodule maps $\varphi:\mathcal{B}\to\mathcal{B}$, it follows that

$$(1.6)\qquad \mathcal{R}\otimes_h{}^\sigma\mathcal{R} \cong \mathcal{M}_{\mathcal{R}'}(\mathcal{B}).$$

Using left and right multiplication to identify $\mathcal{R}\otimes\mathcal{R}$ and $\mathcal{R}'\otimes\mathcal{R}'$ with subspaces of $\mathcal{M}(\mathcal{B})$, the completely bounded maps $\varphi:\mathcal{B}\to\mathcal{B}$, and letting $^-$ and ^{c} denote the relative weak* closure and commutant in the latter space, we may rewrite (1.6):

$$(1.7)\qquad (\mathcal{R}'\otimes\mathcal{R}')^c = (\mathcal{R}\otimes\mathcal{R})^-,$$

i.e., we have the equality of the weak* closure with the "double commutant". (1.6) was used in [5] to identify completely bounded expectations onto $\mathcal{R}'$ with reduced (or "completely bounded") normal virtual diagonals. With (1.7) one may approximate a conditional expectation $E:\mathcal{B}\to\mathcal{R}$ by finite rank completely bounded maps. Haagerup used a completely positive version of this density result to prove directly that injectivity implies Property P [7]. His approximation results may be regarded as bilinear generalizations of the Kaplansky Density Theorem.

For general n, one would expect from (1.4) the relation

(1.8) $$\mathcal{R}^n{}_* \cong(?)\ H^*\bar{\otimes}_{\mathcal{R}'}\mathcal{B}\bar{\otimes}_{\mathcal{R}'}\dots\bar{\otimes}_{\mathcal{R}'}\mathcal{B}\bar{\otimes}_{\mathcal{R}'}H,\quad (n-1 \text{ copies of } \mathcal{B}=\mathcal{B}(H)),$$

and turning to the more general result (1.3), that for any Hilbert space K,

(1.9) $$\mathcal{M}^\sigma(\mathcal{R}^n,\mathcal{B}(K)) \cong(?)\ \mathcal{B}(H,K)\bar{\otimes}_{\mathcal{R}'}\mathcal{B}\bar{\otimes}_{\mathcal{R}'}\dots\bar{\otimes}_{\mathcal{R}'}\mathcal{B}\bar{\otimes}_{\mathcal{R}'}\mathcal{B}(K,H).$$

Indeed what is presumably the key algebraic ingredient for these equations is true (see Proposition 5.1). The difficulty is that we do not know how to give a satisfactory definition for tensor products of the form $\mathcal{B}\bar{\otimes}_{\mathcal{R}'}\dots\bar{\otimes}_{\mathcal{R}'}\mathcal{B}$. In [4] this problem was avoided (for n=3) by using (1.9) to <u>define</u> the mid portion of (1.8). Indeed setting

$$\mathcal{B}\tilde{\otimes}_{\mathcal{R}'}\dots\tilde{\otimes}_{\mathcal{R}'}\mathcal{B} = \mathcal{M}^\sigma(\mathcal{R}^{n-2},\mathcal{B}),\qquad (n-1 \text{ copies of } \mathcal{B}=\mathcal{B}(H))$$

($n\geq 3$), we do have that

(1.10) $$\mathcal{R}^n{}_* \cong H^*\bar{\otimes}_{\mathcal{R}'}\mathcal{B}\tilde{\otimes}_{\mathcal{R}'}\dots\tilde{\otimes}_{\mathcal{R}'}\mathcal{B}\bar{\otimes}_{\mathcal{R}'}H.$$

Taking duals, the corresponding "Multilinear Double Commutant Theorem" then assumes the form

(1.11) $$\mathcal{M}_{\mathcal{R}'}(\mathcal{B}\tilde{\otimes}_{\mathcal{R}'}\dots\tilde{\otimes}_{\mathcal{R}'}\mathcal{B},\mathcal{B}) \cong \mathcal{R}\otimes_h{}^\sigma\mathcal{R}\otimes_h{}^\sigma\dots\otimes_h{}^\sigma\mathcal{R},$$

(n-1 copies of $\mathcal{B}=\mathcal{B}(H)$).

We prove (1.10) and (1.11) for n=3, the case considered in [4] (the general case follows in the same manner). The module approach again provides a much simpler proof. We have also taken the opportunity to correct an omission in that paper.

This research was carried out at the Mathematics Institute at the University of Warwick. The authors wish to thank David Evans and his colleagues for their warm hospitality which made this collaboration possible. This research was partially supported by the National Science Foundation (U.S.A.) and by CNPq-Brazil.

2. The single variable theory

Given a Hilbert space H and a vector $\xi\in H$, we define ξ^* in the dual Hilbert space H^* by $\langle\xi^*,\eta\rangle=\eta\cdot\xi$. We recall that the <u>projective tensor product</u> $H^*\hat{\otimes}H$ is the Banach space completion of the algebraic tensor product $H^*\otimes H$ with the projective norm

$$\|\theta\| = \inf\left\{\sum_{j=1}^{n}\|\xi_j\|\|\eta_j\|:\ \theta=\sum_{j=1}^{n}\xi_j^*\otimes\eta_j\right\}.$$

An infinite sum of the form

$$\theta = \sum_{j=1}^{\infty}\xi_j^*\otimes\eta_j, \tag{2.1}$$

where $\sum\|\xi_j\|\,\|\eta_j\|<\infty$, is absolutely convergent and thus convergent. Conversely any element of $H^*\hat{\otimes}H$ may be written in this form. To prove this (we are following [13], §III.6.4) we choose $\theta_k\in H^*\otimes H$ $(k\geq 1)$ with $\|\theta-\theta_k\|\leq k^{-2}$, and we let $\theta_0=0$. Then letting $\nu_k=\theta_k-\theta_{k-1}$ $(k\geq 1)$, we have $\|\nu_k\|<2k^{-2}$. We may choose ξ_j,η_j $(1\leq j<n_1)$ elements of H with $\theta_1=\sum_{j<n_1}\xi_j^*\otimes\eta_j$, and then $\xi_j,\eta_j\in H$ $(n_k\leq j<n_{k+1})$ with

$$\nu_k=\sum_{n_k\leq j<n_{k+1}}\xi_j^*\otimes\eta_j,$$

$$\sum_{n_k\leq j<n_{k+1}}\|\xi_j\|\|\eta_j\| < 2k^{-2}.$$

It is then evident that θ satisfies (2.1).

We may identify $(H^\infty)^*$ with $(H^*)^\infty$. Given $\xi^*=(\xi_j^*)\in H^{\infty *}$ and $\eta=(\eta_j)\in H^\infty$, we have that

$$\sum\|\xi_j\|\|\eta_j\| \leq \left[\sum\|\xi_j\|^2\right]^{1/2}\left[\sum\|\eta_j\|^2\right]^{1/2} = \|\xi\|\|\eta\|,$$

and thus $\sum\xi_j^*\otimes\eta_j$ converges to an element, which we denote by $\xi^*\odot\eta$. Conversely every element of $H^*\hat{\otimes}H$ may be written in this form. To see this we may assume that $\theta=\sum\lambda_j\xi_j^*\otimes\eta_j$ where $\|\xi_j\|=1$, $\|\eta_j\|=1$, and the λ_j satisfy $\lambda_j\geq 0$, and $\sum\lambda_j<\infty$. One then has that $\theta=\bar{\xi}^*\odot\bar{\eta}$, where $\bar{\xi}=(\lambda_j^{1/2}\xi_j)$ and $\bar{\eta}=(\lambda_j^{1/2}\eta_j)$. In general given any elements $\xi,\eta\in H^\infty$, we have that

$$\|\xi^*\odot\eta\| \leq \|\xi\|\,\|\eta\|, \tag{2.2}$$

and it is not difficult to verify that

$$\|\theta\| = \inf\left\{\|\xi\|\|\eta\|:\ \theta = \xi^*\odot\eta\right\}$$

Let us suppose that $\mathcal{R}\subseteq\mathcal{B}(H)$ is a von Neumann algebra. H is then a left $\mathcal{R}$-module, and we may regard H^* as a right $\mathcal{R}$-module by letting

$$\langle\xi^*x,\eta\rangle = \langle\xi^*,x\eta\rangle,$$

or equivalently, $\xi^*x = (x^*\xi)^*$. We define the $\mathcal{R}$ module projective tensor product $H^*\hat{\otimes}_{\mathcal{R}}H$ by

$$H^*\hat{\otimes}_{\mathcal{R}}H = H^*\hat{\otimes}H/J(\mathcal{R}). \tag{2.3}$$

where

$$J(\mathcal{R}) = \text{closed linear span}\left\{\xi^*\otimes r\eta - \xi^*r\otimes\eta : \xi^*\in H^*, \eta\in H, r\in\mathcal{R}\right\}.$$

Identifying $\mathcal{B}(H)$ with the dual of $H^*\hat{\otimes}H$, we have that

(2.4) $J(\mathcal{R})^{\perp} = \mathcal{R}'$

since $r' \in \mathcal{R}'$ if and only if for all $r \in \mathcal{R}$, $\xi, \eta \in H$,

$$0 = r'r\eta \cdot \xi - rr'\eta \cdot \xi = \langle r', \xi^* \otimes r\eta - \xi^* r \otimes \eta \rangle.$$

It follows that

(2.5) $\mathcal{R}' \cong \left[H^*\hat{\otimes}H/J(\mathcal{R})\right]^* = (H^*\hat{\otimes}_{\mathcal{R}}H)^*.$

We identify the operators $T \in \mathcal{B}(H^\infty)$ with infinite matrices $[T_{ij}]$, $T_{ij} \in \mathcal{B}(H)$, and we use the notation $\mathcal{B}(H^\infty) = \mathbb{M}_\infty(\mathcal{B}(H))$. Given a von Neumann subalgebra $\mathcal{R} \subseteq \mathcal{B}(H)$, we let $\mathbb{M}_\infty(\mathcal{R})$ denote the subalgebra of matrices $r = [r_{ij}]$ with $r_{ij} \in \mathcal{R}$. We regard H^∞ and $H^{\infty *}$ as left and right $\mathbb{M}_\infty(\mathcal{R})$ modules, respectively.

Lemma 2.1: Given $\xi, \eta \in H^\infty$, and $r \in \mathbb{M}_\infty(\mathcal{R})$, we have that

(2.6) $\xi^* \odot r\eta - \xi^* r \odot \eta \in J(\mathcal{R})$.

Proof: Let E_n be the projection of H^∞ onto the subspace H^n of sequences (η_j) with $\eta_j = 0$ for $j > n$. A simple algebraic computation shows that (2.6) is valid if $\eta, \xi \in H^n$ and $E_n r E_n = r$. The general case follows since letting $\eta_n = E_n\eta$, $\xi_n = E_n\xi$ and $r_n = E_n r E_n$, we have that $\|r_n\eta - r\eta\| \to 0$ and $\|r_n{}^*\xi - r^*\xi\| \to 0$, and thus from (2.2),

$$\xi^* \odot r\eta - \xi^* r \odot \eta = \lim \xi_n{}^* \odot r_n\eta_n - \xi_n{}^* r_n \odot \eta_n.$$

We say that a $*$-subalgebra $\mathcal{R} \subseteq \mathcal{B}(H)$ is non-degenerate if $\mathcal{R}H$ is dense in H. It follows that for each $\eta \in H$, $\eta \in (\mathcal{R}\eta)^-$ (see [3]). It is easy to see that if $\mathcal{R}$ is non-degenerate, then $\mathcal{R} \otimes I \subseteq \mathcal{B}(H^\infty)$ is also non-degenerate.

Theorem 2.2: Suppose that $\mathcal{R}$ is a non-degenerate $*$-subalgebra of $\mathcal{B}(H)$. Then $\mathcal{R}_\perp = J(\mathcal{R}')$, and thus

$$\mathcal{R}^- = (\mathcal{R}_\perp)^\perp = J(\mathcal{R}')^\perp = \mathcal{R}''.$$

In particular, if $\mathcal{R}$ is weakly closed, then

$$\mathcal{R}_* \cong H^*\hat{\otimes}_{\mathcal{R}'}H$$

Proof: Given $r' \in \mathcal{R}'$ we have that $rr'\eta \cdot \xi = r\eta \cdot r'^*\xi$, i.e.,

$$\langle r, \xi^* \otimes r'\eta - \xi^* r' \otimes \eta \rangle = 0,$$

and $J(\mathcal{R}') \subseteq \mathcal{R}_\perp$. Conversely let us suppose that $\theta \in \mathcal{R}_\perp$. From above we may assume that $\theta = \xi^* \odot \eta$ with $\xi, \eta \in H^\infty$. We have that for all $r \in \mathcal{R}$,

$$0 = \theta(r) = (r \otimes I)\eta \cdot \xi.$$

Letting $e' = e'^* \in (\mathcal{R} \otimes I)' = \mathbb{M}_\infty(\mathcal{R}')$ be the projection onto $[(\mathcal{R} \otimes I)\eta]^-$, we conclude that e' satisfies $e'\eta=\eta$ and $\xi^*e'=(e'\xi)^*=0$. Thus from Lemma 2.1 we have that

$$\xi^*\odot\eta = \xi^*\odot e'\eta - \xi^*e'\odot\eta \in J(\mathcal{R}').$$

If $\mathcal{R}$ is weakly closed, then the last conclusion follows upon replacing $\mathcal{R}$ by $\mathcal{R}'$ in (2.5).

3. The Bilinear Double Commutant Theorem

Returning to the notation of §1, we begin with a discussion of the predual $\mathcal{M}_*$ of $\mathcal{M}=\mathcal{M}(\mathcal{B},\mathcal{B})$. In previous articles (see, e.g., [5],[8]), $\mathcal{M}_*$ has been identified with the "matricial tensor product" $\mathcal{B}\otimes_{\mathcal{M}}\mathcal{B}_*$. In this paper, we shall instead take advantage of the fact that $\mathcal{B}_*$ may be written as a tensor product of Hilbert spaces. It follows that $\mathcal{M}_*$ may also be described as a suitable completion $H^*\bar{\otimes}\mathcal{B}\bar{\otimes}H$ of the algebraic tensor product $H^*\otimes\mathcal{B}\otimes H$. The definition of the appropriate tensor product norm is reminiscent of the Haagerup norm for operator spaces. [In fact one can regard H and H^* as operator spaces by using the identifications $H\cong\mathcal{B}(\mathbb{C},H)$ and $H^*\cong\mathcal{B}(H,\mathbb{C})$. However, the appropriate "L^1-matricial" norm on $\mathbb{M}_n(H^*\bar{\otimes}\mathcal{B}\bar{\otimes}H)$ is given by the natural identification

$$\mathbb{M}_n(H^*\bar{\otimes}\mathcal{B}(H)\bar{\otimes}H)\cong H^{n*}\bar{\otimes}\mathcal{B}(H^n)\bar{\otimes}H^n,$$

rather than that determined by the Haagerup matrix norms [5].]

Given $T=[T_{ij}]\in\mathbb{M}_n(\mathcal{B})$ and $\xi,\eta\in H^n$ $(n<\infty)$, we define $\xi^*\odot T\odot\eta\in H^*\otimes\mathcal{B}\otimes H$ by

$$\xi^*\odot T\odot\eta = \sum\xi_i^*\otimes T_{ij}\otimes\eta_j.$$

We define the norm of an element $F\in H^*\otimes\mathcal{B}\otimes H$ by

$$\|F\| = \inf\left\{\sum\|\xi^k\|\|T^k\|\|\eta^k\| : F = \sum\xi^{k*}\odot T^k\odot\eta^k\right\},$$

where the sums and matrices are finite. In fact, only one summand is needed, i.e., $\|F\| = |||F|||$, where

$$|||F||| = \inf\left\{\|\xi\|\|T\|\|\eta\| : F = \xi^*\odot T\odot\eta\right\}.$$

To see this it suffices to prove that $|||\ |||$ is subadditive, since then given $\varepsilon>0$ and $F = \sum\xi^{k*}\odot T^k\odot\eta^k$ with

$$\sum\|\xi^k\|\|T^k\|\|\eta^k\| < \|F\| + \varepsilon$$

we have that

$$|||F||| \le |||\sum\xi^{k*}\odot T^k\odot\eta^k||| \le \sum|||\xi^{k*}\odot T^k\odot\eta^k||| \le \sum\|\xi^{k*}\|\|T^k\|\|\eta^k\| < \|F\|+\varepsilon.$$

We recall that for non-negative reals A and B,

$$A^{1/2}B^{1/2} = \inf\left\{\tfrac{1}{2}(\lambda A + \lambda^{-1}B) : \lambda>0\right\}.$$

Given $\varepsilon>0$ and $F_j = \xi^{j*}\odot T^j\odot\eta^j$ $(j=1,2)$, with $\|\xi^j\|\|T^j\|\|\eta^j\| < |||F_j|||+\varepsilon$, we may assume that $\|T^j\|=1$. Then selecting $\lambda_j>0$ with

$$\tfrac{1}{2}(\lambda_j\|\xi^j\|^2 + \lambda_j^{-1}\|\eta^j\|^2) < |||F_j|||+\varepsilon,$$

and letting $\xi = \lambda_1^{1/2}\xi_1\oplus\lambda_2^{1/2}\xi_2$, $T=T_1\oplus T_2$, and $\eta=\lambda_1^{-1/2}\eta_1\oplus\lambda_2^{-1/2}\eta_2$ we have that $F_1+F_2=\xi^*\odot T\odot\eta$, where $\|T\| = 1$. Thus

$$\begin{aligned}\|\xi\|\|T\|\|\eta\| &\le \tfrac{1}{2}\left[\|\xi\|^2 + \|\eta\|^2\right]\\ &\le \tfrac{1}{2}\left[\lambda_1\|\xi_1\|^2 + \lambda_2\|\xi_2\|^2+\lambda_1^{-1}\|\eta_1\|^2 + \lambda_2^{-1}\|\eta_2\|^2\right]\\ &\le |||F_1||| + |||F_2||| + 2\varepsilon,\end{aligned}$$

and $|||F_1+F_2||| \le |||F_1||| + |||F_2|||$.

Consider the pairing

$$\mathcal{M}(\mathcal{B})\times(H^*\otimes\mathcal{B}\otimes H) \to \mathbb{C} \tag{3.1}$$

defined by

$$\langle\varphi,\xi^*\otimes T\otimes\eta\rangle = \varphi(T)\eta\cdot\xi.$$

We have that

$$\begin{aligned}\|\varphi\|_{cb} &= \sup\left\{|\varphi_n(T)\eta\cdot\xi| : T\in\mathbb{M}_n(\mathcal{B}); \eta,\xi\in H^n;\ \|T\|=\|\xi\|=\|\eta\|=1, n\in\mathbb{N}\right\}\\ &= \sup\left\{|\langle\varphi,\ \xi^*\odot T\odot\eta\rangle| :\ \|T\|=\|\xi\|=\|\eta\|=1,\ n\in\mathbb{N}\right\}.\end{aligned} \tag{3.2}$$

It follows that in general,

$$|\langle\varphi,F\rangle| \le \|\varphi\|\|F\|.$$

This may be used to show that $\|\ \|$ is a norm on $H^*\otimes\mathcal{B}\otimes H$. Given

$$F = \sum\xi_j^*\otimes T_j\otimes\eta_j \ne 0,$$

we may assume that the T_j are linearly independent, $\zeta_1^*\otimes T_1\otimes\eta_1\ne 0$, and that ξ_1,η_1 are unit vectors. We define $S\in\mathcal{B}$ by $S\zeta = (\zeta\cdot\eta_1)\xi_1$, and we choose $\beta\in\mathcal{B}^*$ satisfying $\beta(T_j)=\delta_{1j}$. Then letting $\varphi\in\mathcal{M}(\mathcal{B})$ be the map $\varphi(T)=\beta(T)S$, we have that $1= \langle\varphi,F\rangle \le \|\varphi\|\|F\|$, and thus $\|F\|\ne 0$.

We let $H^*\bar{\otimes}\mathcal{B}\bar{\otimes}H$ denote the completion of $H^*\otimes\mathcal{B}\otimes H$. It follows from (3.2) that the pairing (3.1) induces an isometry of $\mathcal{M}(\mathcal{B})$ into $\left[H^*\bar{\otimes}\mathcal{B}\bar{\otimes}H\right]^*$. On the other hand, given $F\in\left[H^*\bar{\otimes}\mathcal{B}\bar{\otimes}H\right]^*$, we may define a map $\varphi:\mathcal{B}\to\mathcal{B}$ by letting

$$\varphi(T)\eta\cdot\xi = F(\xi^*\otimes T\otimes\eta).$$

Fixing T, the usual arguments show that this formula indeed uniquely determines an element of $\mathcal{B}$. Given $\xi,\eta\in H^\infty$, and $T\in\mathcal{B}(H^\infty)$,

$$|\varphi_\infty(T)\eta\cdot\xi| = |F(\xi^*\odot T\odot\eta)| \le \|F\|\ \|\xi^*\odot T\odot\eta\| \le \|F\|\|\xi\|\|T\|\|\eta\|,$$

hence $\|\varphi\|_{cb}\le\|F\|$. Since φ determines F, the isometry is surjective, i.e.,

$$\mathcal{M}(\mathcal{B}) \cong \left[H^*\bar{\otimes}\mathcal{B}\bar{\otimes}H\right]^*.$$

Given $\xi,\eta\in H^\infty$ and $T\in\mathcal{B}(H^\infty)$ with $\|T\|\le 1$, we define $\xi_n=E_n\xi$, $\eta_n=E_n\eta$, and $T_n=E_nTE_n$ (see §2). If $m>n$, then

$$\xi_n^* \odot T_n \odot \eta_n = \xi_n^* \odot T_m \odot \eta_n,$$

and

$$\|\xi_m^* \odot T_m \odot \eta_m - \xi_n^* \odot T_n \odot \eta_n\| \leq \|\xi_m - \xi_n\| \|\eta\| + \|\xi\| \|\eta_m - \eta_n\|,$$

and it follows that $\xi_n^* \odot T_n \odot \eta_n$ is norm convergent. We denote the limit by $\xi^* \odot T \odot \eta$. Conversely, every element of $H^* \bar{\otimes} \mathcal{B} \bar{\otimes} H$ has this form, as may be seen by imitating the corresponding argument for $H^* \hat{\otimes} H$. Given $\varphi \in \mathcal{M}(\mathcal{B})$, we may define φ_∞: $\mathcal{B}(H^\infty) \to \mathcal{B}(H^\infty)$ by the strong limit

$$\varphi_\infty(T) = \lim \varphi_n(E_n T E_n),$$

or equivalently,

$$\varphi_\infty(T)\xi \cdot \eta = \langle \varphi, \xi^* \odot T \odot \eta \rangle$$

for $\xi, \eta \in H^\infty$. It is easy to check that $\varphi_\infty(T)$ has the matrix $[\varphi(T_{ij})]$ (see[5]).

We have a natural injection $\kappa: \mathcal{B} \otimes \mathcal{B} \to \mathcal{M}(\mathcal{B})$ defined by

$$\kappa(A \otimes B)(T) = ATB.$$

Although this is not a multiplicative isomorphism (we have deliberately avoided using $\mathcal{B}^{op}$), the image is a subalgebra of $\mathcal{M}(\mathcal{B})$, and the same is true for $\mathcal{R} \otimes \mathcal{R}$ for any subalgebra $\mathcal{R} \subseteq \mathcal{B}$. In this section we shall notationally identify $\mathcal{B} \otimes \mathcal{B}$ with its image $\kappa(\mathcal{B} \otimes \mathcal{B})$. We write $\mathcal{S}^c$ for the commutant of a subset $\mathcal{S} \subseteq \mathcal{M}(\mathcal{B})$. In particular, we have that

$$(\mathcal{R} \otimes \mathcal{R})^c = \mathcal{M}_{\mathcal{R}}(\mathcal{B}),$$

where $\mathcal{M}_{\mathcal{R}}(\mathcal{B})$ denotes the $\mathcal{R}$-bimodule maps $\varphi: \mathcal{B}(H) \to \mathcal{B}(H)$.

Given a non-degenerate *-algebra $\mathcal{R} \subseteq \mathcal{B}(H)$, we may regard H^*, $\mathcal{B}(H)$, and H as right, two sided, and left $\mathcal{R}$ modules, respectively. We define the $\mathcal{R}$-tensor product by

$$H^* \bar{\otimes}_{\mathcal{R}} \mathcal{B} \bar{\otimes}_{\mathcal{R}} H = H^* \bar{\otimes} \mathcal{B} \bar{\otimes} H / J(\mathcal{R} \otimes \mathcal{R}),$$

where

$$J(\mathcal{R} \otimes \mathcal{R}) = \text{closed span} \left\{ \xi^* \otimes rTs \otimes \eta - \xi^* r \otimes T \otimes s\eta : \xi^* \in H^*, \eta \in H, r, s \in \mathcal{R} \right\}.$$

It is evident that $\mathcal{M}_{\mathcal{R}}(\mathcal{B})$ is just the annihilator $J(\mathcal{R} \otimes \mathcal{R})^\perp$, and thus we have that

$$(3.3) \qquad \mathcal{M}_{\mathcal{R}}(\mathcal{B}) = \left[H^* \bar{\otimes}_{\mathcal{R}} \mathcal{B} \bar{\otimes}_{\mathcal{R}} H \right]^*.$$

The following result is more difficult to prove than its analogue Lemma 2.1. The argument used there does not apply since we need not have that $\xi_n^* \odot r_n T s_n \odot \eta_n$ converges in norm to $\xi^* \odot rTs \odot \eta$. Instead we appeal to a beautiful result of G. May (based in part on a suggestion of E. Neuhardt) [10].

Lemma 3.1: Given $\xi,\eta\in H^\infty$, $T\in\mathcal{B}(H^\infty)$, and $r,s\in\mathbb{M}_\infty(\mathcal{R})$, we have that

$$\xi^*\odot rTs\odot\eta - \xi^* r\odot T\odot s\eta\in J(\mathcal{R}\otimes\mathcal{R}).$$

Proof: It suffices to prove that if $\varphi\in J(\mathcal{R}\otimes\mathcal{R})^\perp$, then φ annilhilates the given element, i.e., that if φ is an $\mathcal{R}$ bimodule map, then φ_∞ is an $\mathbb{M}_\infty(\mathcal{R})$ bimodule map. Despite the fact that φ is not assumed to be normal, this is the case (see [10], or [6], Corollary 4.3.)

Theorem 3.2: Suppose that $\mathcal{R}$ is a non-degenerate *-subalgebra of $\mathcal{B}(H)$. Then $(\mathcal{R}\otimes\mathcal{R})_\perp = J(\mathcal{R}'\otimes\mathcal{R}')$, and thus

$$(\mathcal{R}\otimes\mathcal{R})^- = (\mathcal{R}'\otimes\mathcal{R}')^c = \mathcal{M}_{\mathcal{R}'}(\mathcal{B}).$$

Proof: Given that $\omega\in(\mathcal{R}\otimes\mathcal{R})_\perp$, we may assume that $\omega=\xi^*\odot T\odot\eta$. Then for all $r,s\in\mathcal{R}$, we have that

$$\omega(r\otimes s) = (r\otimes I)T(s\otimes I)\eta\cdot\xi=0.$$

Letting e' and f' be the projections on $[(\mathcal{R}\otimes I)\xi]^-$ and $[(\mathcal{R}\otimes I)\eta]^-$, respectively, we have $\xi^*e'=(e'\xi)^*=\xi^*$, $f'\eta=\eta$, and $e'Tf'=0$. Thus

$$\omega = \xi^*e'\odot T'\odot f'\eta - \xi^*\odot e'T'f'\odot\eta,$$

and from Lemma 3.1, $\omega\in J(\mathcal{R}'\otimes\mathcal{R}')$. The reverse inclusion is trivial.

Let us suppose that $\mathcal{R}\subseteq\mathcal{B}(H)$ is a von Neumann algebra. From the Christensen-Sinclair representation (1.4), the function

$$\rho: H^*\bar{\otimes}\mathcal{B}\bar{\otimes}H \to \mathcal{R}^2{}_*$$

defined by

$$\rho(\xi^*\odot T\odot\eta)(r,s) = (r\otimes I)T(s\otimes I)\eta\cdot\xi,$$

is a surjective quotient map. It is obvious that the kernel of this map is precisely $(\mathcal{R}\otimes\mathcal{R})_\perp$. Since the latter coincides with $J(\mathcal{R}'\otimes\mathcal{R}')$, we conclude that

$$\mathcal{R}^2{}_* \cong H^*\bar{\otimes}_{\mathcal{R}'}\mathcal{B}\bar{\otimes}_{\mathcal{R}'}H, \tag{3.4}$$

and thus from (3.3),

Corollary 3.3 [5]: Suppose that $\mathcal{R}\subseteq\mathcal{B}(H)$ is a von Neumann algebra. Then

$$\mathcal{R}\otimes^\sigma{}_h\mathcal{R} \cong \mathcal{M}_{\mathcal{R}'}(\mathcal{B}).$$

§4. The Trilinear Case

In this section we shall prove (1.10) and (1.11) for the case n=3. We recall from §1 that by definition, $\mathcal{B}\tilde{\otimes}_{\mathcal{R}}\mathcal{B} = \mathcal{M}^\sigma(\mathcal{R}',\mathcal{B})$. We shall take the liberty of writing $\varphi = S \odot T$ for operators of the form $\varphi(r') = S(r' \otimes I)T$. Given $1 \leq k \leq \infty$, we may identify $\mathbb{M}_k(\mathcal{B} \tilde{\otimes}_{\mathcal{R}} \mathcal{B})$ with $\mathcal{M}^\sigma(\mathcal{R}', \mathbb{M}_k(\mathcal{B}))$, where the corresponding elements have the form $\varphi(r') = S(r' \otimes I)T$, $S \in \mathcal{B}(H^k, H^J)$, $T \in \mathcal{B}(H^J, H^k)$ for a suitable index set J. We may regard this space as an $\mathbb{M}_k(\mathcal{B})$ bimodule. It is a simple matter to verify that given $B \in \mathbb{M}_k(\mathcal{B})$,

$$\|\varphi \oplus \psi\| = \max\{\|\varphi\|, \|\psi\|\}, \quad \|B\varphi\| \leq \|B\|\|\varphi\|, \quad \|\varphi B\| \leq \|\varphi\|\|B\|,$$

and in particular, $\mathcal{B} \tilde{\otimes}_{\mathcal{R}} \mathcal{B}$ is an abstract operator space in the sense of Ruan [12] (we are indebted to him for this remark). We define $\mathbb{M}_\infty(\mathcal{B} \tilde{\otimes}_{\mathcal{R}} \mathcal{B})$ to be the $\infty \times \infty$ matrices $v = [v_{ij}]$, $v_{ij} \in \mathcal{B} \tilde{\otimes}_{\mathcal{R}} \mathcal{B}$ for which there is a constant C with

$$\|[v_{ij}]_{i,j \leq n}\| \leq C$$

for all n. If one uses Ruan's theory [12] to realize $\mathcal{B} \tilde{\otimes}_{\mathcal{R}} \mathcal{B}$ as a linear space of operators $\mathcal{V}$ on a Hilbert space L, then it is easy to see that these correspond to the operators $v = [v_{ij}] \in \mathcal{B}(L^\infty)$ with $v_{ij} \in \mathcal{V}$. More generally, given a completely bounded map $\Phi: \mathcal{B} \tilde{\otimes}_{\mathcal{R}} \mathcal{B} \to \mathcal{B}(K)$, we may define a bounded map $\Phi_\infty: \mathbb{M}_\infty(\mathcal{B} \tilde{\otimes}_{\mathcal{R}} \mathcal{B}) \to \mathbb{M}_\infty(\mathcal{B}(K))$ by

$$\Phi_\infty(v) = \text{strong} \lim \Phi_n(E_n v E_n) = [\Phi(v_{ij})].$$

We define the tensor product norm as in §3, i.e., given F in the algebraic tensor product $H^* \otimes [\mathcal{B} \tilde{\otimes}_{\mathcal{R}} \mathcal{B}] \otimes H$, we let

$$\|F\| = \inf\{\|\xi\|\|\varphi\|\|\eta\| : F = \xi^* \odot \varphi \odot \eta, \xi, \eta \in H^k, \varphi \in \mathbb{M}_k(\mathcal{B} \tilde{\otimes}_{\mathcal{R}} \mathcal{B}), 1 \leq k < \infty\},$$

where we use the same convention regarding $\odot$, and we let $H^* \bar{\otimes} [\mathcal{B} \tilde{\otimes}_{\mathcal{R}} \mathcal{B}] \bar{\otimes} H$ be the corresponding completion. Once again each element of the completion has the form $F = \xi^* \odot \varphi \odot \eta$, where $\xi, \eta \in H^\infty$, $\varphi \in \mathbb{M}_\infty(\mathcal{B} \tilde{\otimes}_{\mathcal{R}} \mathcal{B})$.

We have a pairing

$$\mathcal{M}(\mathcal{B} \tilde{\otimes}_{\mathcal{R}} \mathcal{B}, \mathcal{B}) \times (H^* \bar{\otimes} [\mathcal{B} \tilde{\otimes}_{\mathcal{R}} \mathcal{B}] \bar{\otimes} H) \to \mathbb{C} \tag{4.1}$$

defined by

$$\langle \Phi, \xi^* \odot \varphi \odot \eta \rangle = \Phi_\infty(\varphi)\eta \cdot \xi,$$

or letting $\varphi = S \odot T$,

$$\langle \Phi, \xi^* \odot S \odot T \odot \eta \rangle = \Phi_\infty(S \odot T)\eta \cdot \xi.$$

From this duality it is easily seen that

$$\mathcal{M}(\mathcal{B} \tilde{\otimes}_{\mathcal{R}} \mathcal{B}, \mathcal{B}) \cong [H^* \bar{\otimes} [\mathcal{B} \tilde{\otimes}_{\mathcal{R}} \mathcal{B}] \bar{\otimes} H]^*.$$

We define the $\mathcal{R}$ bimodule tensor product by

$$H^* \bar{\otimes}_{\mathcal{R}} [\mathcal{B} \tilde{\otimes}_{\mathcal{R}} \mathcal{B}] \bar{\otimes}_{\mathcal{R}} H = H^* \bar{\otimes} [\mathcal{B} \tilde{\otimes}_{\mathcal{R}} \mathcal{B}] \bar{\otimes} H / J(\mathcal{R} \otimes \mathcal{R}),$$

where

$$J(\mathcal{R}\otimes\mathcal{R}) = \text{closed span}\left\{\xi^*\otimes r\varphi s\otimes\eta - \xi^* r\otimes\varphi\otimes s\eta : \varphi\in \mathcal{B}\tilde{\otimes}_{\mathcal{R}}\mathcal{B},\ \xi,\eta\in H, r,s\in\mathcal{R}\right\}.$$

It is evident that the annihilator $J(\mathcal{R}\otimes\mathcal{R})^{\perp}$ of $J(\mathcal{R}\otimes\mathcal{R})$ in $\mathcal{M}(\mathcal{B}\tilde{\otimes}_{\mathcal{R}}\mathcal{B},\mathcal{B})$ is just the subspace $\mathcal{M}_{\mathcal{R}}(\mathcal{B}\tilde{\otimes}_{\mathcal{R}}\mathcal{B},\mathcal{B})$ of $\mathcal{R}$-bimodule maps, and thus we have that

(4.1) $$\mathcal{M}_{\mathcal{R}}(\mathcal{B}\tilde{\otimes}_{\mathcal{R}}\mathcal{B},\mathcal{B}) \cong \left[H^*\bar{\otimes}_{\mathcal{R}}[\mathcal{B}\tilde{\otimes}_{\mathcal{R}}\mathcal{B}]\bar{\otimes}_{\mathcal{R}}H\right]^*.$$

We have the following analogue of May's result (Lemma 3.1):

Lemma 4.1: Given $\xi,\eta\in H^{\infty},\varphi\in \mathbb{M}_{\infty}(\mathcal{B}\tilde{\otimes}_{\mathcal{R}}\mathcal{B})$, and $r,s\in\mathbb{M}_{\infty}(\mathcal{R})$, we have that

$$\xi^*\odot r\varphi s\odot\eta - \xi^* r\odot\varphi\odot s\eta\in J(\mathcal{R}\otimes\mathcal{R}).$$

Proof: It suffices to prove that if $\Phi\in J(\mathcal{R}\otimes\mathcal{R})^{\perp}\subseteq\mathcal{M}(\mathcal{B}\tilde{\otimes}_{\mathcal{R}}\mathcal{B},\mathcal{B})$, then Φ annihilates the given element, i.e., that if

$$\Phi: \mathcal{B}\tilde{\otimes}_{\mathcal{R}}\mathcal{B} = \mathcal{M}^{\sigma}(\mathcal{R}',\mathcal{B}) \to \mathcal{B}$$

is an $\mathcal{R}$ bimodule map, then

$$\Phi_{\infty}:\mathbb{M}_{\infty}(\mathcal{B}\tilde{\otimes}_{\mathcal{R}}\mathcal{B}) \to \mathbb{M}_{\infty}(\mathcal{B})$$

is an $\mathbb{M}_{\infty}(\mathcal{R})$ bimodule map. This is proved in [6], Corollary 4.3.

Let us suppose that $\mathcal{R}$ is a subalgebra of $\mathcal{B}(H)$. Given $r,s\in\mathcal{R}$, we have a corresponding map

$$\kappa:\mathcal{R}\otimes\mathcal{R}\otimes\mathcal{R} \to \mathcal{M}(\mathcal{B}\tilde{\otimes}_{\mathcal{R}'}\mathcal{B},\mathcal{B})$$

defined by letting

$$\kappa(r\otimes s\otimes t)(\varphi) = r\varphi(s)t.$$

Notationally identifying $\mathcal{R}\otimes\mathcal{R}\otimes\mathcal{R}$ with its image, we have that $(\mathcal{R}\otimes\mathcal{R}\otimes\mathcal{R})_{\perp}$ is a subset of the predual $H^*\bar{\otimes}[\mathcal{B}\tilde{\otimes}_{\mathcal{R}'}\mathcal{B}]\bar{\otimes}H$.

Theorem 4.2: Suppose that $\mathcal{R}\subseteq\mathcal{B}(H)$ is a non-degenerate *-subalgebra of $\mathcal{B}(H)$. Then $(\mathcal{R}\otimes\mathcal{R}\otimes\mathcal{R})_{\perp} = J(\mathcal{R}'\otimes\mathcal{R}')$, and thus

$$(\mathcal{R}\otimes\mathcal{R}\otimes\mathcal{R})^{-} = \mathcal{M}_{\mathcal{R}'}(\mathcal{B}\tilde{\otimes}_{\mathcal{R}'}\mathcal{B},\mathcal{B}).$$

Proof: Given that $\Omega\in(\mathcal{R}\otimes\mathcal{R}\otimes\mathcal{R})_{\perp}$, let us suppose that $\Omega=\xi^*\odot\varphi\odot\eta$ where $\xi,\eta\in H^{\infty}$, and $\varphi\in\mathcal{M}^{\sigma}(\mathcal{R},\mathbb{M}_{\infty}(\mathcal{B}))$. Then for all $r,s,t\in\mathcal{R}$ we have that

$$\Omega(r,s,t) = (r\otimes I)\varphi(s)(t\otimes I)\eta\cdot\xi = 0.$$

Letting e' and f' be the projections on $[(\mathcal{R}\otimes I)\xi]^{-}$ and $[(\mathcal{R}\otimes I)\eta]^{-}$, respectively, we have $\xi^*e'=(e'\xi)^*=\xi^*$, $f'\xi=\xi$, and $e'\varphi f'=0$. Thus

$$\Omega = \xi^*e'\odot\varphi\odot f'\eta - \xi^*\odot e'\varphi f'\odot\eta,$$

and from Lemma 4.1, $\Omega\in J(\mathcal{R}\otimes\mathcal{R})$. The reverse inclusion is trivial.

Once again we have that the Christensen-Sinclair representation (1.4) implies that the map

$$\rho:\ H^*\bar{\otimes}\left[\mathcal{B}\tilde{\otimes}_{\mathcal{R}'}\mathcal{B}\right]\bar{\otimes}H \to \mathcal{R}^3{}_*$$

defined by

$$\rho(\xi^*\odot\varphi\odot\eta)(r,s,t) = r\varphi(s)t\eta\cdot\xi$$

is a surjective quotient map. It has kernel is $(\mathcal{R}\times\mathcal{R})_\perp$, and since the latter coincides with $J(\mathcal{R}'\otimes\mathcal{R}')$,

$$\mathcal{R}^3{}_* \cong H^*\bar{\otimes}_{\mathcal{R}'}\left[\mathcal{B}\tilde{\otimes}_{\mathcal{R}'}\mathcal{B}\right]\bar{\otimes}_{\mathcal{R}'}H.$$

Replacing $\mathcal{R}$ by $\mathcal{R}'$ in (4.1),

Corollary 4.3 [4]: Suppose that $\mathcal{R}\subseteq\mathcal{B}(H)$ is a von Neumann algebra. Then

$$\mathcal{R}\otimes^\sigma{}_h\mathcal{R}\otimes^\sigma{}_h\mathcal{R} \cong \mathcal{M}_{\mathcal{R}'}(\mathcal{B}\tilde{\otimes}_{\mathcal{R}'}\mathcal{B},\mathcal{B}).$$

The proof given in [4] is incomplete. In that paper, the author restricted his attention to elements in $H^*\bar{\otimes}\left[\mathcal{B}\tilde{\otimes}_{\mathcal{R}'}\mathcal{B}\right]\bar{\otimes}H$ of the form $\xi^*\odot\varphi\odot\eta$ with $\xi,\eta\in H^n$ and $\varphi\in \mathbb{M}_n(\mathcal{B}\tilde{\otimes}_{\mathcal{R}'}\mathcal{B})$ for finite n. This was done in order to avoid proving Lemma 4.1 (Corollary 4.3 in [6] was unavailable at that time). The difficulty occurs at the very end of the proof, in which it is asserted that the functional M_φ is bounded. Although it is true that any element F of the dense subspace $H^*\otimes\left[\mathcal{B}\tilde{\otimes}_{\mathcal{R}'}\mathcal{B}\right]\otimes H$ may be written in the form the form $\xi^*\odot T\odot\eta$ for finite rank vectors and matrices, it is not immediately evident that one can also require that $\|F\|$ be close to $\|\xi^*\|\|T\|\|\eta\|$ for such vectors and matrices.

5. The general case

The algebraic result proved below provides some evidence in favor of the conjectured relation (1.10). Any diagram D of bridging maps (1.2) determines a function $F=F_D\in\mathcal{M}^\sigma(\mathcal{R}^n,\mathcal{B}(K))$ by

$$F(r_1,\ldots,r_n) = S_0(r_1\otimes I)S_1\ldots S_{n-1}(r_n\otimes I)S_n.$$

Fixing a sufficiently large index set J, we may assume that $J_1=\ldots=J_n=J$. We write $S_0\odot\ldots\odot S_n$ for the n+1-tuple

$$(S_n,\ldots,S_0)\in\mathcal{B}(H^J,K)\times\mathcal{B}(H^J)\times\ldots\times\mathcal{B}(H^J)\times\mathcal{B}(K,H^J).$$

We define an equivalence relation $\equiv$ on these objects by letting

$$S_0\odot\ldots\odot S_kr'\odot S_{k+1}\odot\ldots\odot S_n \equiv S_0\odot\ldots\odot S_k\odot r'S_{k+1}\odot\ldots\odot S_n,$$

for arbitrary k and $r'\in\mathcal{R}'$, and then saturating to an equivalence.

Proposition 5.1: Suppose that D is a diagram of bridging maps given by (1.2). Then if $F_D = 0$, it follows that

$$S_0 \odot \ldots \odot S_n \equiv 0.$$

Proof: By assumption we have that for all $r_k \in \mathcal{R}$,

$$S_0(r_1 \otimes I)S_1 \ldots S_{n-2}(r_{n-1} \otimes I)S_{n-1}(r_n \otimes I)S_n = 0.$$

Letting $e_n' \in (\mathcal{R} \otimes I)' = \mathcal{R}' \otimes \mathcal{B}(H^J)$ be the projection on $[(\mathcal{R} \otimes I)S_n K]^-$, it follows that $S_n = e_n' S_n$, and

$$S_0(r_1 \otimes I)S_1 \ldots S_{n-2}(r_{n-1} \otimes I)S_{n-1}e_n' = 0$$

for all $r_1, \ldots, r_{n-1} \in \mathcal{R}$. Similarly letting $e_{n-1}' \in (\mathcal{R} \otimes I)'$ be the projection on $[(\mathcal{R} \otimes I)S_{n-1}e_n' H^J]^-$, we find that $S_{n-1}e_n' = e_{n-1}' S_n e_n'$, and

$$S_0(r_1 \otimes I)S_1 \ldots S_{n-2}e_{n-1}' = 0.$$

Continuing in this fashion, we obtain a projection e_1' which in particular satisfies $S_0 e_1' = 0$. We conclude that

$$\begin{aligned} S_0 \odot \ldots S_{n-1} \odot S_n &= S_0 \odot \ldots \odot S_{n-2} \odot S_{n-1} \odot e_n' S_n \\ &\equiv S_0 \odot \ldots \odot S_{n-2} \odot S_{n-1} e_n' \odot S_n \\ &= S_0 \odot \ldots \odot S_{n-2} \odot e_{n-1}' S_{n-1} e_n' \odot S_n \\ &\equiv S_0 \odot \ldots \odot S_{n-2} e_{n-1}' \odot S_{n-1} e_n' \odot S_n \\ &\equiv \ldots \\ &\equiv S_0 e_1' \odot \ldots \odot S_{n-2} e_{n-1}' \odot S_{n-1} e_n' \odot S_n \\ &= 0. \end{aligned}$$

The formulation of an appropriate tensor product in (1.9) will require using the weak operator topology, since given $S \in \mathbb{M}_{1\infty}(\mathcal{B}(H))$, $T \in \mathbb{M}_{\infty 1}(\mathcal{B}(H))$, we cannot expect expressions of the form

$$(S \odot T)_{ij} = \sum S_{ik} \otimes T_{kj}$$

to converge in a norm topology on $\mathcal{B}(H) \bar{\otimes} \mathcal{B}(H)$. Presumably one must provide the latter space with both a norm and a weak topology. We shall not pursue this question here.

References

1. E. Christensen, E. Effros, and A. Sinclair, Completely bounded multilinear maps and C*-algebraic cohomology, Inv. Math., to appear.

2. E. Christensen and A. Sinclair, Representations of completely bounded multilinear operators, J. Fnal. Anal. 72 (1987), 151-181.

3. J. Dixmier, Les Algebres d'Opérateurs dans l'Espace Hilbertien, 2nd ed., Gauthier-Villars, Paris, 1969.

4. E. Effros, On multilinear completely bounded maps, Contemp. Math. 62 (1987),479-501.

5. _________ and A. Kishimoto, Module maps and Hochschild-Johnson cohomology, Ind. Math. J. 36 (1987), 257-276.

6. ________ and Z.-J. Ruan, Representations of operator bimodules and their applications, to appear.

7. U. Haagerup, A new proof of the equivalence of injectivity and property P, Jan. 1984, upub. ms.

8. T. Itoh, Completely bounded maps and a certain cross norm, to appear.

9. R.V. Kadison, A generalized Schwarz inequality and algebraic invariants for C*-algebras, Ann. Math. 56 (1952), 494-503.

10. G. May, Das geordnete normale Haagerup-Tensorprodukt einer von Neumann-Algebra und seine Anwedung, Diplomarbeit, Fach. Mathematik der Universität des Saarlandes, 1986.

11. V. Paulsen, Completely Bounded Maps and Dilations, Pitman Research Notes in Math., No. 146, Longman Scientific & Technical, New York, 1986.

12. Z. Ruan, Subspaces of C*-algebras, J. Fnal. Anal., to appear.

13. H. H. Schaeffer, Topological Vector Spaces, Graduate Texts in Math., No. 3, Springer-Verlag, New York, 1970.

E. Effros
Mathematics Department
UCLA
Los Angeles, CA 90024
USA

Ruy Exel
Departamento de Matematica
IMEUSP
Caixa Postal 20570 (Ag. Iguatemi)
01498 San Paulo SP
Brazil

LOOP SPACES, CYCLIC HOMOLOGY AND THE CHERN CHARACTER

Ezra Getzler
Mathematics Department, Harvard University, Cambridge, MA 02138, USA.

John D.S. Jones
Mathematics Institute, The University of Warwick, Coventry CV4 7AL, England.

Scott B. Petrack
Département de Mathématiques, Université de Paris-Sud, Bâtiment 425, 91405 Orsay, France.

INTRODUCTION The purpose of this note is to show how ideas from cyclic homology may be used to study geometrical and analytic properties of loop spaces. As such, the material in this note is not directly related to operator algebras; on the other hand it is not too distant either since cyclic homology is one of several good things to come out of the study of operator algebras in recent years.

Geometry and analysis on loop spaces is a subject of much current interest. One of the ideas which has emerged is the following, rather striking, point of view on the Chern character and the index theorem. Let X be a smooth compact oriented manifold with no boundary and let LX be the space of all smooth loops in X. This loop space is an infinite dimensional manifold, modelled on a Fréchet space. The circle T acts on LX by rotating loops; this action is smooth and the fixed point set is the manifold X regarded as the space of constant loops. We are principally interested in properties of the smooth T-manifold LX. The relation between the Chern character, the index theorem and loop spaces is given by the following facts. Let E be a complex vector bundle on X equipped with a connection ∇.

(a) There is a T-invariant, equivariantly closed differential form $\omega(E,\nabla)$ on LX with the property that the restriction of $\omega(E,\nabla)$ to X, the space of constant loops, is precisely the usual Chern character form, $\mathrm{ch}(E,\nabla) = \mathrm{Trace}(e^F)$ where F is the curvature of ∇. This form $\omega(E,\nabla)$ is Bismut's equivariant extension of the Chern character.

(b) Suppose that X is a spin manifold. Then there is a T-invariant, equivariantly closed current μ on LX with the property that $<\mu, \omega(E,\nabla)> = \text{index } D_E$, where D_E is the Dirac operator on X coupled to the vector bundle E using the connection ∇. This is Witten's current on the loop space.

These ideas evolved as follows. The initial idea, which is due to Witten is that if X is a spin manifold, then there is an invariant, equivariantly closed current μ on LX such that $<\mu,1> = \text{index } D$, where D is the Dirac operator on X. Some related ideas are described in [W]. This idea was developed, very beautifully, by Atiyah in [A]. In [B], Bismut studied the twisted Dirac operator D_E and concluded that there was an invariant equivariantly closed form $\omega(E,\nabla)$ on LX such that $<\mu, \omega(E,\nabla)> = \text{index } D_E$. Up to this point most of the arguments were formal, rather than rigorous, and involved "renormalisation" by dividing by infinite constants but it is quite clear that they give a new point of view on the index theorem and the Chern character. In [B], Bismut went on to rigorously construct the form $\omega(E,\nabla)$.

We will approach these constructions from a rather different point of view; the idea is to exploit the relation between cyclic homology and loop spaces. The precise relation between cyclic homology and T-equivariant cohomology is described in full generality in [J]; here we will concentrate on one special case which we will describe in terms most suited to our applications.

Let $\Omega(X)$ be the differential graded algebra of differential forms on X. We will explain how to construct dual maps

cyclic cycles over $\Omega(X)$ $\rightarrow$ invariant, equivariantly closed forms on LX

invariant, equivariantly closed currents on LX $\rightarrow$ cyclic cocycles over $\Omega(X)$.

Here, by cyclic cycle or cocycle we mean a cycle in the appropriate form of Connes' b, B double complex. This opens up the possibility of constructing Bismut's form ω as a cyclic cycle and interpreting Witten's current μ as a cyclic cocycle.

In this note we will explain how to construct ω as a cyclic cycle; there is a subtlety however in that it will be constructed as a cyclic cycle over the algebra $\Omega(X)\otimes\Omega_T(T)$ where $\Omega_T(T)$ is the space of forms on T which are invariant under rotations. We only give a summary of some of our results and methods; in particular very few detailed proofs are given. A fuller account, including a discussion of the cyclic cocycle determined by Witten's current will be given in forthcoming publications.

§1 ITERATED INTEGRALS The relation between cyclic cycles and cocycles, on the one hand, and differential forms and currents, on the other, is given by Chen's iterated integrals [Ch2]. Let ω be a 1-form on X, then ω defines an obvious function on LX

$$\gamma \to \int_\gamma \omega.$$

This elementary construction can be generalized to a method of using forms on X to construct forms on LX.

Loops will be parametrized by the unit interval [0,1] in **R**. Let ω be a differential form on X and define $\omega(t)$ to be the differential form $e_t^*(\omega)$ where $e_t : LX \to X$ is evaluation at t. Let $\iota : \Omega(LX) \to \Omega(LX)$ denote interior product with the vector field generating the circle action on LX. We will denote by Δ^k the k-simplex consisting of points

$$\{ (t_1, \dots, t_k) \mid t_i \in \mathbf{R}, 0 \le t_1 \le t_2 \le \dots \le t_k \le 1 \}.$$

Given differential forms $\omega_0, \dots, \omega_k$ on X, their **invariant iterated integral** is the differential form on LX defined by the formula

$$\rho_A(\omega_0, \dots, \omega_k) = \int_{I\times\Delta^k} \omega_0(s) \wedge \iota\omega_1(t_1) \wedge \dots \wedge \iota\omega_k(t_k)\, ds\, dt_1 \dots dt_k$$

where, of course, we take $s+t_i$ mod 1. This form is invariant under the action of the circle and has degree given by

$$\deg(\rho_A(\omega_0, \dots, \omega_k)) = \deg(\omega_0) + \dots + \deg(\omega_k) - k.$$

There are two basic operators on forms which occur in the definition of equivariant cohomology, the exterior derivative d and the interior product ι. We look for formulas for these operators applied to forms given by iterated integrals.

There are two basic operators in cyclic homology, the Hochschild boundary operator b and Connes' coboundary operator B. The formulas for these operators are as follows:

$$b(\omega_0, \dots, \omega_k) = \sum_{i=0}^{k} (-1)^i (\omega_0, \dots, \omega_i\omega_{i+1}, \dots, \omega_k) + (-1)^\varepsilon (\omega_k\omega_0, \omega_1, \dots, \omega_{k-1})$$

where $\varepsilon = k + |\omega_k| (|\omega_0| + \dots + |\omega_{k-1}|)$ so that $(-1)^\varepsilon$ is the natural sign of the cyclic permutation acting on k graded objects, and

$$B(\omega_0, \dots, \omega_k) = \sum_{i=0}^{k} (-1)^{\varepsilon_i} (1, \omega_i, \dots, \omega_k, \omega_0, \dots, \omega_{i-1}).$$

where $\varepsilon_i = i + (|\omega_0| + \dots + |\omega_{i-1}|) (|\omega_i| + \dots + |\omega_k|)$. We will continue to denote by d the usual extension of the exterior derivative to (k+1)-tuples of differential forms, that is

$$d(\omega_0, \dots, \omega_k) = \sum_{i=0}^{k} (-1)^{|\omega_0| + \dots + |\omega_{i-1}|} (\omega_0, \dots, \omega_{i-1}, d\omega_i, \omega_{i+1}, \dots, \omega_k)$$

Theorem (a) $d\rho_A(\omega_0, \dots, \omega_k) = \rho_A d(\omega_0, \dots, \omega_k) + (-1)^q \rho_A b(\omega_0, \dots, \omega_k)$

where $q = \deg(\omega_0) + \dots + \deg(\omega_k)$.

(b) $\iota\rho_A(\omega_0, \dots, \omega_k) = \rho_A B(\omega_0, \dots, \omega_k)$.

The proof of this theorem is elementary but it will not be given here. The theorem shows that the operators b, B of cyclic homology correspond, under iterated integrals, to the operators d, ι of equivariant cohomology and what follows is an effort to exploit this rather striking fact.

§2 EQUIVARIANT COHOMOLOGY We now give a rapid account of one or two of the key features of equivariant cohomology. Suppose that Y is a smooth manifold equipped with an action of the circle. If Y is infinite dimensional, then by differential forms we will mean smooth differential forms in the sense appropriate to manifolds modelled on a Hilbert space, a Banach space or a Fréchet space as the case may be. We use d for the exterior derivative and ι for the interior product with the vector field generating the circle action. Now make the following definitions.

$\Omega^q_T(Y)$ is the space of T invariant q-forms on Y

$$\Omega_T(Y) = \prod_q \Omega^q_T(Y)\ ,\ \Omega^{ev}_T(Y) = \prod_q \Omega^{2q}_T(Y)\ ,\ \Omega^{odd}_T(Y) = \prod_q \Omega^{2q+1}_T(Y)$$

Define $d_T : \Omega_T(Y) \to \Omega_T(Y)$ to be the operator $d_T = d + \iota$. Then since $\iota^2 = 0$ and $d\iota + \iota d = 0$ on the space of invariant forms it follows that $d_T d_T = 0$ and so we may form cohomology groups; by definition

$$H^{ev}_T(Y) = \frac{\ker(d_T : \Omega^{ev}_T(Y) \to \Omega^{odd}_T(Y))}{\operatorname{im}(d_T : \Omega^{odd}_T(Y) \to \Omega^{ev}_T(Y))}$$

$$H^{odd}_T(Y) = \frac{\ker(d_T : \Omega^{odd}_T(Y) \to \Omega^{ev}_T(Y))}{\operatorname{im}(d_T : \Omega^{ev}_T(Y) \to \Omega^{odd}_T(Y))}\ .$$

Note that if the action of T on Y is trivial then it follows immediately that

$$H^{ev}_T(Y) = \prod_q H^{2q}(Y) = H^{ev}(Y)$$

$$H^{odd}_T(Y) = \prod_q H^{2q+1}(Y) = H^{odd}(Y)$$

where $H^q(Y)$ means the usual deRham cohomology of Y. The following theorem is proved in [JP].

Theorem The inclusion $X \to LX$ of the fixed point set of the circle action on LX induces isomorphisms

$$H^{ev}_T(LX) \to H^{ev}_T(X) = H^{ev}(X)\ ,\ H^{odd}_T(LX) \to H^{odd}_T(X) = H^{odd}(X)$$

In particular this theorem shows that there is an equivariant cohomology class in $H_T{}^{ev}(LX)$ whose restriction to X is the Chern character of E ; this is the cohomology class which corresponds to $ch(E) \in H^{ev}(X)$ under the above isomorphism. However we also know that if E comes equipped with extra structure, a connection ∇, then we can find a canonical closed differential form $ch(E,\nabla)$ whose cohomology class is ch(E); this is a very important fact. Naturally, we would like a canonical representative for the corresponding equivariant cohomology class on LX and Bismut's invariant, equivariantly closed form $\omega(E,\nabla)$ is precisely such a representative.

There are various other equivariant cohomology groups which we should mention. First we can replace the direct product of invariant forms by the direct sum and define groups

$$\Omega_{T,f}(Y) = \sum_q \Omega_T^q(Y)$$

$$H_{T,f}^{ev}(Y)\ ,\ H_{T,f}^{odd}(Y)$$

The subscript f stands for finite. Of course if Y is finite dimensional then $\Omega_{T,f}(Y)$ is equal to $\Omega_T(Y)$ and the two forms of equivariant cohomology coincide. However if Y is infinite dimensional then this is not the case as the following theorem of Goodwillie, [G1], demonstrates very dramatically.

Goodwillie's Theorem If X is a compact, finite dimensional, simply connected, smooth manifold then

$$H_{T,f}^{ev}(LX) = H_{T,f}^{odd}(LX) = 0.$$

This theorem is highly relevant to the construction of invariant, equivariantly closed forms on LX whose restriction to X represents the Chern character of E; it proves that any such form must involve an infinite number of terms.

The two forms of equivariant cohomology we have discussed so far are both variations of the usual equivariant cohomology. The standard form of equivariant cohomology is constructed as follows, compare [AB]. Define

$$\Omega_T(Y)^{(p)} = \sum_{0 \le 2i \le p} \Omega_T^{p-2i}(Y)$$

There is a cochain complex

$$\dots \to \Omega_T(Y)^{(p)} \to \Omega_T(Y)^{(p+1)} \to \dots$$

with boundary maps d_T and the cohomology groups $H_T{}^p(Y)$ of this cochain complex are the usual equivariant cohomology groups. There are natural maps

$$\dots \to H_T^p(Y) \to H_T^{p+2}(Y) \to \dots$$

and it is easy to prove that the direct limit of this sequence is

$$H^{ev}_{T,f}(Y) \text{ if p is even} \quad , \quad H^{odd}_{T,f}(Y) \text{ if p is odd}$$

It is not clear whether $H_T^{ev}(Y)$ and $H_T^{odd}(Y)$ depend only on the groups $H_T^{p}(Y)$.

The final point to make about the equivariant cohomology of infinite dimensional manifolds is that it seems reasonable to study the cohomology groups which arise when $\Omega_T(Y)$ is replaced by the set of sequences $(\omega_0, \omega_1, \ldots, \omega_n, \ldots)$ of forms, where ω_n satisfies some prescribed decay conditions as n tends to infinity.

§3 CYCLIC HOMOLOGY Next we discuss cyclic homology. Let Ω be a differential graded algebra, with unit, over **C**. We assume that Ω is graded so that the differential increases degree by one. First we describe the Hochschild complex of Ω. Define

$$C(\Omega) = \prod_p \Omega \otimes (I\Omega)^{\otimes p}$$

where $I\Omega = \Omega/\mathbf{C}$ with **C** embedded in Ω as scalar multiples of the identity. We give $C(\Omega)$ the differential

$$\partial = d \pm b$$

where the sign is $(-1)^q$ on an element $\omega_0 \otimes \ldots \otimes \omega_k$ with $q = \deg(\omega_0) + \ldots + \deg(\omega_k)$. The Hochschild complex is graded so that

$$\deg(\omega_0 \otimes \ldots \otimes \omega_k) = \deg(\omega_0) + \ldots + \deg(\omega_k) - k,$$

therefore, if $C^k(\Omega)$ denotes the space of homogeneous elements of degree k, then

$$\partial : C^k(\Omega) \to C^{k+1}(\Omega).$$

As in §2, we will use the notation $C^{ev}(\Omega)$ and $C^{odd}(\Omega)$ for the even and odd parts of $C(\Omega)$ and the notation $C_f(\Omega)$ for the sub-complex obtained by replacing the direct product, in the definition of $C(\Omega)$, by the direct sum.

Connes' B operator gives a map

$$B : C^k(\Omega) \to C^{k-1}(\Omega)$$

such that $BB = \partial B + B\partial = 0$. Therefore, in the definition of equivariant cohomology we may replace $\Omega_T(Y)$ by $C(\Omega)$, d by ∂ and ι by B to define cohomology groups

$$H^{ev}(C(\Omega)) \ , \ H^{odd}(C(\Omega)) \ , \ H_f^{ev}(\Omega) \ , \ H_f^{odd}(C(\Omega)) \ , \ H^p(C(\Omega)) \, .$$

These are various cyclic homology groups of the algebra Ω;

$$H^p(C(\Omega)) = HC^-_{-p}(\Omega)$$

where the HC^- groups are the "negative" cyclic homology groups described in [G2], [HJ] and [J],

$$H_f^{ev}(C(\Omega)) = HC_{ev,f}(\Omega) \ , \ H_f^{odd}(C(\Omega)) = HC_{odd,f}(\Omega)$$

are the even and odd parts of periodic cyclic homology. The groups $HC_{ev}(\Omega) = H^{ev}(C(\Omega))$ and $HC_{odd}(\Omega) = H^{odd}(C(\Omega))$ are completed versions of periodic cyclic homology.

We should point out that some of our notation for cyclic homology groups and equivariant cohomology groups is non-standard so care should be taken when translating between our notation and the notation used by other authors.

Using the theorem in §1, ρ_A defines a map of complexes

$$(C(\Omega(X)), \partial + B) \to (\Omega_T(LX), d + \iota).$$

By definition a cyclic cycle over $\Omega(X)$ is a cycle for $\partial + B$ whereas an invariant, equivariantly closed form on LX is a cycle for $d + \iota$, so we obtain the map

cyclic cycles over $\Omega(X)$ $\to$ invariant, equivariantly closed forms on LX

mentioned in the introduction. The dual map of complexes gives the map

invariant, equivariantly closed currents on LX $\to$ cyclic cocycles over $\Omega(X)$.

Suppose that X is simply connected, compact and finite dimensional then the map induced by ρ_A in cohomology gives isomorphisms

$$H^p(C(\Omega X)) \cong H_T^p(LX) \ , \ H_f^{ev}(C(\Omega X)) \cong H_{T,f}^{ev}(LX) \ , \ H_f^{odd}(C(\Omega X)) \cong H_{T,f}^{odd}(LX) \, .$$

This is proved in [J] using general topological considerations; an explicit proof which exploits special features of differential forms and the geometry of LX will be given in a future paper. The question of whether the induced maps

$$H^{ev}(C(\Omega X)) \to H_T^{ev}(LX) \ , \ H^{odd}(C(\Omega X)) \to H_T^{odd}(LX)$$

are isomorphisms is quite delicate and will not be discussed here.

§4 THE CHERN CHARACTER Now we turn to the construction of the Chern character as a cyclic cycle. We will in fact work over the algebra $\Omega = \Omega(X) \otimes \Omega_T(T) = \Omega_T(X \times T)$. Here the circle acts on itself by multiplication and on $X \times T$ by the product of the trivial action on X and the multiplication action on T. The general element of Ω is of the form

$$\omega = \alpha + \beta \, dt$$

where α and β are differential forms on X and dt is the standard invariant 1-form on T. This algebra Ω will be given the equivariant differential d_T; explicitly

$$d_T (\alpha + \beta \, dt) = d\alpha + (-1)^{|\beta|} \beta + d\beta \, dt.$$

If $\omega_0 , \dots , \omega_k$ are elements of Ω their iterated integral is a differential form on $L(X \times T)$ and since the forms ω_i are invariant under the action of the circle on $X \times T$ the form $\rho_A(\omega_0 , \dots , \omega_k)$ is invariant under the action of the circle on $L(X \times T)$ given by the combination of rotating loops and the action on $X \times T$; from now on we will always equip $L(X \times T)$ with this action. There is a natural equivariant map

$$j : LX \to L(X \times T)$$

defined by $j(\gamma)(t) = (\gamma(t), t)$ and we will denote by σ the composite homomorphism

$$C(\Omega) \to \Omega_T(L(X \times T)) \to \Omega_T(LX)$$

given by $\sigma = j^* \rho_A$. There is an explicit formula for σ,

$$\sigma(\omega_0, \dots , \omega_k) = \int_{I \times \Delta^k} \alpha_0(s) \wedge (\iota\alpha_1(s+t_1) + \beta_1(s+t_1)) \wedge \dots \wedge (\iota\alpha_k(s+t_k) + \beta_k)) \, ds \, dt_1 \dots dt_k$$

where $\omega_i = \alpha_i + \beta_i \, dt$, and, as in §1, $s+t_i$ is taken modulo 1.

Now let E be a vector bundle on X with a connection ∇; then we will construct an element ψ in $C(\Omega)$ with the property that $\sigma(\psi)$ is an invariant, equivariantly closed form on LX whose restriction to X is $ch(E,\nabla)$. To construct ψ we will assume that the bundle E is given by an idempotent $e \in M_n(C^\infty(X))$ in the sense that

$$E \subseteq X \times \mathbf{C}^n \ , \ E_x = e(x)\mathbf{C}^n.$$

We will also assume that the connection ∇ is the Grassmannian connection on E; precisely this means that if s is a section of E, regarded as a vector valued function on X, then the covariant derivative of s is given by the formula

$$\nabla s = eds.$$

Every bundle with connection can be constructed in this way; this corresponds to the fact that the Grassmannian connection on the universal bundle over the Grassmann manifold is universal, compare [NR]. However, distinct idempotents may give rise to the same bundle with connection. Our construction depends on the choice of e. One way to view the construction is as a construction on the Grassmannian connection on the universal bundle over the Grassmann manifold.

Let $f = 1 - e$, and let F be the bundle defined by f. Define

$$A = ede + fdf,$$

so A is a matrix of 1-forms on X. In fact A is the connection form of the direct sum of the Grassmannian connection on E with the Grassmannian connection on F. Precisely

$$(d + A)(es) = ed(es) \ , \ (d + A)(fs) = fd(fs)$$

where s is any $\mathbf{C}^n$-valued function on X, so that $d + A$ applied to a section of E is the covariant derivative and $d + A$ applied to a section of F is also the covariant derivative. Now define

$$R = dA + A \wedge A = dede,$$

so R is the curvature of the connection $d + A$ on the trivial bundle $X \times \mathbf{C}^n$, and also define

$$N = A - Rdt$$

It is straightforward to verify the following formulas

$$\text{(4.1)} \qquad \begin{aligned} d_T e &= [e, N] \\ d_T N + N \wedge N &= 0. \end{aligned}$$

Define $N_k \in C(\Omega)$ by

$$N_k = e \otimes N \otimes \dots \otimes N$$

where there are k factors N. An easy computation, using 4.1, will establish the following formula;

(4.2) $$d_T N_{k-1} = bN_k.$$

So we can construct a cycle for ∂, which will be called θ, from the N_k by the following formula

$$\theta = \sum_{k=0}^{\infty} \varepsilon_k N_k$$

where the signs ε_k are chosen as follows

$$\varepsilon_k = 1 \text{ if } k = 0, 3 \text{ modulo } 4$$

$$\varepsilon_k = -1 \text{ if } k = 1, 2 \text{ modulo } 4.$$

Now we apply the generalized trace map

$$tr_k : M_n(\Omega)^{\otimes(k+1)} \to \Omega^{\otimes(k+1)}.$$

This is defined by identifying $M_n(\Omega)$ with $M_n(C) \otimes \Omega$ and then using the formula

$$tr_k(m_0 \otimes \omega_0 \otimes m_1 \otimes \omega_1 \otimes \ldots \otimes m_k \otimes \omega_k) = \text{Trace}(m_0 m_1 \ldots m_k)\, \omega_0 \otimes \omega_1 \otimes \ldots \otimes \omega_k.$$

Define ψ as follows;

$$\psi = \sum_{k=0}^{\infty} \varepsilon_k tr_k(N_k) \in C(\Omega).$$

From 4.2 we deduce that

$$\partial\psi = 0,$$

but note that

$$B\psi \neq 0$$

so that ψ is not a cycle in $C(\Omega)$. However in the formula for $B\psi$ each of the terms occuring in the sum involves an element of Ω^0 in a place other then the first. But it is clear from the formula for σ that $\sigma(\omega_0, \ldots, \omega_k) = 0$ if $\omega_i \in \Omega^0$ for some i with $1 \leq i \leq k$ and therefore $\sigma(B\psi) = 0$. This shows that

$$(d + \iota)\, \sigma(\psi) = \sigma((\partial + B)\, \psi) = 0$$

and therefore $\sigma(\psi)$ is an invariant, equivariantly closed differential form on LX.

The slightly unsatisfactory point that $(\partial + B)\psi \neq 0$ may be tidied up as follows. The problem is that we have not "normalised" $C(\Omega)$ properly; there is a subcomplex D of "degenerate" elements with the following properties:

(a) $\omega_0 \otimes ... \otimes \omega_k \in D$ if $\omega_i \in \Omega^0$ for some i with $1 \leq i \leq k$.

(b) D is acyclic with respect to $\partial + B$.

(c) $\sigma(D) = 0$.

We should really work in the "normalised complex" $C(\Omega)/D$. We will not go into the details of this refinement here, it is explained precisely in [Ch1]. In plain terms it means that we may modify ψ by an element in the kernel of σ to obtain a cycle for $\partial + B$.

Now we prove that $\sigma(\psi)$ gives $ch(E,\nabla)$ when restricted to X. This follows directly from the formula for $\sigma(\psi)$

$$\sum_{k=0}^{\infty} \varepsilon_k \int_{I \times \Delta^k} tr_k\left(e(s)(\iota A(s+t_1) - R(s+t_1)) \wedge ... \wedge (\iota A(s+t_k) - R(s+t_k))\right) \, ds\, dt_1 ... dt_k$$

When we restrict to the space of constant loops then, since ι is defined by taking the interior product with a vector field which vanishes on the space of constant loops, the terms $\iota A(t)$ become zero; evidently $e(t)$ becomes e and $R(t)$ becomes $R = dede$. So the whole expression gives

$$\sum_{k=0}^{\infty} (-1)^k \varepsilon_k \int_{I \times \Delta^k} tr\,(e(dede)^k) = \sum_{k=0}^{\infty} \frac{(-1)^k \varepsilon_k}{k!} tr\,(e(dede)^k)$$

which, up to constants, is the standard formula for $ch(E,\nabla)$ in terms of e.

We will study the properties of ψ in more detail in forthcoming papers. In particular we will compare it with Bismut's $\omega(E,\nabla)$ and examine its dependence on the idempotent e. Finally we give some explanation of why we choose to work over $\Omega_T(X \times T)$. It is not difficult to see that any form on $L(X)$ in the image of ρ_A is invariant, not only under rotations of the loops but under any diffeomorphism of the circle acting on LX by reparametrizing loops. Bismut's form $\omega(E,\nabla)$ does not have this property. We get around this difficulty by first constructing forms on $L(X \times T)$ and then restricting these forms to $L(X)$ by using the map $j : X \to L(X \times T)$. This device breaks the symmetry and we end up with forms which are invariant under rotations but not under any diffeomorphism of the circle.

REFERENCES

[A] M.F. Atiyah, Circular symmetry and stationary phase approximation. Colloque en l'honneur de Laurent Schwartz, *Astérisque* 131 (1985), 43-59.

[AB] M.F. Atiyah and R. Bott, The moment map and equivariant cohomology. *Topology* 23 (1984), 1-28.

[B] J-M. Bismut, Index theorem and equivariant cohomology on the loop space. *Commun. Math. Phys.* 98 (1985), 213-237.

[Ch1] K-T. Chen, Reduced bar constructions on de Rham Complexes. *Algebra, Topology and Category Theory* (edited by A. Heller and M. Tierney), 19-32, Academic Press, (1976).

[Ch2] K-T. Chen, Iterated path integrals. *Bulletin of the AMS* 83 (1977), 831-879.

[Co] A. Connes, Non commutative differential geometry, Parts I and II. *Publ. Math. Inst. Hautes Etudes Sci.* 62, (1985), 41-144.

[G1] T.G. Goodwillie, Cyclic homology, derivations and the free loop space. *Topology* 24 (1985), 187-215.

[G2] T.G. Goodwillie, Relative algebraic K-theory and cyclic homology. *Annals of Math. 124* (1986), 347-402.

[HJ] C.E. Hood and J.D.S. Jones, Some algebraic properties of cyclic homology groups. To appear in: *K-Theory*.

[J] J.D.S. Jones, Cyclic homology and equivariant homology. *Inventiones Math.* 87 (1987), 403-424.

[JP] J.D.S. Jones and S.B. Petrack, Le théorème des points fixes en cohomologie équivariante. To appear in: *Comptes Rendus de l'Académies des Sciences - Série I : Mathématique.*

[NR] M.S. Narasimhan and S. Ramanan, Existence of universal connections. *Amer. J. Math.* 83 (1961), 563-572.

[W] E. Witten, Supersymmetry and Morse theory, *J. Diff. Geom.* 17 (1982), 661-692.

THE WEYL THEOREM AND BLOCK DECOMPOSITIONS

R.V. Kadison
University of Pennsylvania

Dedicated to Marshall H. Stone on his eighty fifth birthday

1 INTRODUCTION

In [9; Satz VI], Weyl proves that each bounded, self-adjoint operator can be "perturbed" by the addition of a self-adjoint compact operator to yield a self-adjoint operator that is diagonalized by some orthonormal basis. Von Neumann [8] gives a simpler proof and sharpens this result. In answer to a question of Halmos [2], Berg [1] (see also [3]) proves the same result for normal operators.

In a sweeping extension of these results, Zsido [10], using techniques developed by Halmos [4], when introducing the important concept of quasitriangularity, and methods of von Neumann algebra theory, proves the corresponding result for countably generated commutative C*-subalgebras of a countably decomposable, infinite, semi-finite factor. Among other results, Zsido shows that each self-adjoint operator in a countably decomposable factor of type I_∞ or II_∞ is the sum of a diagonal operator $\Sigma_{n=1}^{\infty} a_n E_n$ and a self-adjoint operator C in the (unique) proper, norm-closed ideal generated by the finite projections, where each E_n is one-dimensional in the I_∞ case and of trace 1 relative to a given normal, semi-finite tracial weight in the II_∞ case. Kaftal [7] has extended some of these results to include normal operators.

In the next section, we present an extension of the Weyl theorem that is stronger than the Zsido extension, in one way, and weaker in another. The proof is simpler than most arguments that yield the Weyl theorem. It deals with the possibility of special block decompositions of the self-adjoint operators in these factors and makes use of the "block diagonalization" theorem of [5]. Both this result (Theorem A) and the Zsido result yield the Weyl theorem at once in the classical (I_∞) case.

We acknowledge, with gratitude, the partial support of the NSF (USA) and the SERC (GB) during the research for this article.

2 BLOCK DECOMPOSITION AND WEYL'S THEOREM IN SEMI-FINITE FACTORS

By using [5], we show directly that each self-adjoint operator has a block decomposition with blocks of arbitrarily prescribed small dimension, the "off-diagonal" matrix lying in the (unique) proper norm-closed ideal.

Theorem A.

Suppose $\mathcal{M}$ is an infinite countably decomposable, semi-finite factor and $\mathcal{F}$ is the (unique, proper) norm-closed, two-sided ideal, the norm closure of the ideal $\mathcal{I}$ of operators with finite range projection. Let F be a non-zero finite projection and H a self-adjoint operator in $\mathcal{M}$. Then there is an orthogonal family of projections $\{G_1, G_2, \ldots\}$ in $\mathcal{M}$ such that $\Sigma G_j = I$, $G_j \precsim F$ $(j = 1,2,\ldots)$, and $\Sigma_{j\neq k} \; G_j H G_k$ is an operator in $\mathcal{F}$.

Proof.

We use the spectral projections for H and iterated bisection of $[-||H||, ||H||]$ to construct projections $\{E_{jk}\}$, where $j \in \{0,1,2,\ldots\}$ and $k \in \{1,\ldots,2^j\}$, such that $\Sigma_{k=1}^{2^j} E_{jk} = I$. Let $\{E_1, E_2, \ldots\}$ be an orthogonal family of finite projections in $\mathcal{M}$ with sum I. Let F_0 be 0 and F_j be

$$\cup\{R(E_{jk}E_h) : k \in \{1,\ldots,2^j\}, \; h \in \{1,\ldots,j\}\} .$$

Since E_{jk} is the sum of $E_{j+1k'}$ for certain k', we have that $F_j \leq F_{j+1}$.

Let a_{jk} be the midpoint of the interval for which E_{jk} is the spectral projection and let A_j be $\Sigma_{k=1}^{2^j} a_{jk}E_{jk}$. Then $||H-A_j|| \leq 2^{-j}||H||$. Let F_{jk} be

$$\cup\{R(E_{jk}E_h) : h \in \{1,\ldots,j\}\} .$$

Then $F_{jk} \leq E_{jk}$ for each k, $\{F_{jk} : k \in \{1,2,\ldots,2^j\}\}$ is an orthogonal family since $\{E_{jk} : k \in \{1,2,\ldots,2^j\}\}$ is, and $F_j = \Sigma_{k=1}^{2^j} F_{jk}$.

As $E_h = (\Sigma_{k=1}^{2^j} E_{jk})E_h$, we have that $\Sigma_{h=1}^{j} E_h \leq F_j$. Thus $\cup_{j=1}^{\infty} F_j = I$, and $\{F_j\}$ is strong-operator convergent to I. Since $R(E_{jk}E_h) \sim R((E_{jk}E_h)^*) = R(E_hE_{jk}) \leq E_h$ [6; Proposition 6.1.6], and E_h is finite, each $R(E_{jk}E_h)$ is finite. From [6; Theorem 6.3.8], each

F_j is finite. Let M_j be $F_j - F_{j-1}$ $(j = 1,2,\ldots)$. We show that M_{j+1} commutes with A_j. To see this, note that if $j \le j'$, then $E_{j'k} \le E_{jk'}$ for some k', whence $F_{j'k} \le E_{j'k} \le E_{jk'}$. It follows that $F_{j'k}$ commutes with A_j as does $F_{j'}$. Hence M_{j+1} $(=F_{j+1} - F_j)$ commutes with A_j. We conclude that

$$||HM_j - M_jH|| \le ||(H-A_{j-1})M_j|| + ||M_j(A_{j-1}-H)|| \le 2^{-(j-2)}||H|| .$$

Thus

$$\begin{aligned} ||F_j[(HM_j-M_jH)M_j-M_j(HM_j-M_jH)]F_j|| &= ||\sum_{k=1}^{j-1}(M_kHM_j+M_jHM_k)|| \\ &\le 2||HM_j-M_jH|| \\ &\le 2^{-(j-3)}||H|| . \end{aligned}$$

Since $\Sigma_{k=1}^{j-1}(M_kHM_j + M_jHM_k)$ $(=C_j)$ has a finite range projection (from [6; Theorem 6.3.8]) and $||C_j|| \le 2^{-(j-3)}||H||$, we have that $\Sigma_{j=2}^{\infty} C_j$ converges in norm to an operator $\Sigma_{j \ne k} M_jHM_k$ in F.

We can find a finite orthogonal family of equivalent projections with sum M_1 and with each projection in the family equivalent to a subprojection of F. (If $M_1 \precsim F$, we may use $\{M_1\}$ as this family.) Let $\{G_{jk}^{(1)} : j,k = 1,\ldots,n_1\}$ be a self-adjoint system of matrix units with $\{G_{jj}^{(1)}\}$ the orthogonal family (cf. [6; Lemma 6.6.4]). From [5], there is a unitary operator U_1 in M_1MM_1 (acting on $M_1(H)$) such that $U_1M_1HM_1U_1^*$ has a diagonal matrix relative to $\{G_{jk}^{(1)}\}$; that is

$$U_1M_1HM_1U_1^* = \sum_{j=1}^{n_1} G_{jj}^{(1)}\, U_1M_1HM_1U_1^*\, G_{jj}^{(1)} .$$

Equivalently,

$$M_1HM_1 = \sum_{j=1}^{n_1} U_1^*G_{jj}^{(1)}\, U_1M_1HM_1U_1^*\, G_{jj}^{(1)}U_1 .$$

We now repeat this construction for each of $M_2, M_3, \ldots,$ producing unitary elements $U_2, U_3, \ldots,$ in M_2MM_2, $M_3MM_3, \ldots,$ and matrix unit systems $\{G_{jk}^{(2)}\}$, $\{G_{jk}^{(3)}\}, \ldots,$ such that

$$M_hHM_h = \sum_{j=1}^{n_h} U_h^*G_{jj}^{(h)}U_hM_hHM_hU_h^*G_{jj}^{(h)}U_h .$$

With $\mathcal{H}$ the Hilbert space on which $\mathcal{M}$ acts and x a vector in $M_h(\mathcal{H})$, let Ux be U_hx. Then U defines a unitary operator in $\mathcal{M}$ and U commutes with each M_h. Moreover, for h in $\{1,2,\ldots\}$,

$$M_hHM_h = \sum_{j=1}^{n_h} U^*G_{jj}^{(h)}UM_hHM_hU^*G_{jj}^{(h)}U .$$

We enumerate $U^*G_{11}^{(1)}U, U^*G_{22}^{(1)}U,\ldots,U^*G_{n_1n_1}^{(1)}U, U^*G_{11}^{(2)}U,\ldots$ as $G_1,G_2,\ldots$. Thus

$$M_1 = G_1 +\ldots+ G_{n_1}, M_2 = G_{n_1+1} +\ldots+ G_{n_1+n_2}, \ldots .$$

It follows that $\Sigma_{h=1}^{\infty} M_hHM_h = \Sigma_{j=1}^{\infty} G_jHG_j$ and that the compact operator $\Sigma_{j\neq k} M_jHM_k$ coincides with $\Sigma_{j\neq k} G_jHG_k$, for the apparently missing terms G_jHG_k, where $j \neq k$ and G_j,G_k are subprojections of the same M_h, are all 0 since G_j and G_k are distinct principal units of the same matrix unit systems that diagonalize the operators M_hHM_h in these cases. □

Corollary B.

With $\mathcal{H}$ a separable Hilbert space and H a self-adjoint element in $\mathcal{B}(\mathcal{H})$, there is an orthonormal basis for $\mathcal{H}$ relative to which the matrix for H with the diagonal entries replaced by 0 is compact.

Proof.

Let F be a one-dimensional projection in Theorem A. Then each G_j is either 0 or one dimensional. For our orthonormal basis, we choose an orthonormal basis for each $G_j(\mathcal{H})$ and use their union. □

Once we are reconciled to the loss of diagonalizability of self-adjoint "matrices" when we pass from finite to infinite, we note that, at any rate, the spectral theorem allows us to find an orthonormal basis relative to which all off-diagonal terms are small. Indeed, with ε positive and $\{E_1,\ldots,E_n\}$ a finite orthogonal family of spectral projections for H with sum I such that $||H-\Sigma\lambda_jE_j|| < \varepsilon$, where $\lambda_1,\ldots,\lambda_n$ are points of the spectrum of H, the union of orthonormal bases for $E_1(\mathcal{H}),\ldots,E_n(\mathcal{H})$ provides us with an orthonormal basis relative to which the matrix of $H-\Sigma\lambda_j E_j$ has each entry of absolute value less than ε and $\Sigma\lambda_j E_j$ has a diagonal matrix. Thus each off-diagonal entry of H has absolute value less than ε. (In fact, the operator whose matrix has 0 at diagonal entries and the off-diagonal entries of H has norm less than 2ε, for all diagonal entries of $H-\Sigma\lambda_j E_j$ have absolute value less than ε, whence the diagonal matrix with these diagonal entries has norm less than ε as does $H-\Sigma\lambda_j E_j$.) Arranging for smallness of the off-diagonal matrix in the sense of compactness, as in Corollary B, is striking. By starting with M_r, r sufficiently large, in place of M_1 in the proof of Theorem A (so that our series estimates begin with 2^r), we can arrange that the off-diagonal matrix has small norm as well as being compact.

Some simple considerations allow us to draw the matrix representation result of Corollary B directly from the Weyl theorem. If we know that $H = D + C$, where D is a diagonal relative to the orthonormal basis $\{e_n\}$ and C is compact, then $\{||Ce_n||\}$ tends to 0 since $\{e_n\}$ converges weakly to 0, and compact operators convert weakly convergent sequences to norm convergent sequences [6; Exercise 2.8.20]. Thus $\{\langle Ce_n, e_n\rangle\}$ converges to 0. The diagonal entries for the matrix of C relative to $\{e_n\}$ are $\{\langle Ce_n, e_n\rangle\}$. It follows that the diagonal operator with diagonal entries the diagonal of C is compact [6; Exercise 2.8.26]. Hence the difference of C and this diagonal operator is compact. That difference is the off-diagonal matrix obtained from the matrix of H relative to $\{e_n\}$.

With some sharpening of these techniques, we can see that the "block diagonal" matrix formed from C is compact for all "sizes" of blocks. We make this precise in the following proposition.

Proposition C.

If C is a compact operator on the separable Hilbert space $\mathcal{H}$ and $\{E_n\}$ is an orthogonal family of projections, then $\Sigma_n E_n C E_n$ is compact.

Proof.

Let α be the linear mapping of $\mathcal{B}(\mathcal{H})$ into itself that assigns $\Sigma E_n T E_n$ to T. Since

$$||\Sigma E_n T E_n|| = \sup_n\{||E_n T E_n||\} \leq \sup_n\{||E_n||\ ||T||\ ||E_n||\} \leq ||T||,$$

we have that $||\alpha|| \leq 1$. Suppose that F is a one-dimensional projection and x is a unit vector in its range. Then E_nFE_n is self-adjoint with range spanned by E_nx. If $E_nFE_n \neq 0$, then E_nFE_n is a positive multiple of the one-dimensional projection with range spanned by E_nx. If y is a unit vector in the range of E_n, then

$$E_nFE_ny = \langle y,x\rangle E_nx = ||E_nx||^2\langle y,||E_nx||^{-1}E_nx\rangle ||E_nx||^{-1}E_nx ,$$

whence that multiple (and the norm of E_nFE_n) is $||E_nx||^2$. Since $\Sigma||E_nx||^2 < \infty$, we have that $\alpha(F)$ is the norm limit of the operators $\Sigma_{n=1}^{m} E_nFE_n$ $(m = 1,2,\ldots)$, each of which has finite-dimensional range. Thus $\alpha(F)$ is compact.

If T has finite-dimensional range, it is a linear combination of one-dimensional projections (not necessarily mutually orthogonal) and $\alpha(T)$ is compact. Since α is bounded, we have that $\alpha(C)$ $(= \Sigma E_nCE_n)$ is compact when C is compact. □

The argument of Proposition C relies on the fact that $\{||E_nFE_n||\}$ tends to 0. This need not hold in a factor of Type II_∞ when F is replaced by a projection of (relative) dimension 1, even when each E_n has relative dimension 1. To see this, let $F = \Sigma_{n=1}^{\infty} F_n$, where F_n is a subprojection of E_n of dimension 2^{-n}. Then $||E_nFE_n|| = 1$ for all n. Of course $\Sigma E_nFE_n = \Sigma F_n = F$, in this case, so that ΣE_nFE_n is in the unique, proper, norm-closed ideal of the factor. These comments raise some doubts about the validity of the assertion of Proposition C as formulated for (countably decomposable) factors of Type II_∞. By using more sophisticated techniques, we prove that the assertion of Proposition C is valid for all countably decomposable semi-finite factors. (Although Proposition D subsumes Proposition C, it was our purpose to argue the assertion of Proposition C in relatively basic terms.)

<u>Proposition D.</u>

Let M be an infinite, countably decomposable, semi-finite factor, I the two-sided ideal in M consisting of all operators in M whose range projection is finite, and F be the norm closure of I. If $\{E_1,E_2,\ldots\}$ is an orthogonal family of projections in M and $A \in F$, then $\Sigma_{j=1}^{\infty} E_jAE_j \in F$.

Proof.

Suppose, first, that A is a finite projection F in M. Let ρ be the unique, normal, semi-finite tracial weight on M such that $\rho(F) = 1$ (cf. [6; §8.5]). Then

$$\begin{aligned}\rho\left(\sum_{j=1}^{\infty} E_jFE_j\right) &= \sum_{j=1}^{\infty} \rho(E_jFE_j) = \lim_{n\to\infty} \sum_{j=1}^{n} \rho(E_jFE_j)\\ &= \lim_{n\to\infty} \sum_{j=1}^{n} \rho(FE_jF) = \lim_{n\to\infty} \rho\left(F\left(\sum_{j=1}^{n} E_j\right)F\right)\\ &\le \lim_{n\to\infty} \rho(F) = \rho(F) .\end{aligned}$$

As noted in [6; Proposition 8.5.1], M_ρ is a two-sided ideal. Since M is infinite and semi-finite, $I \notin M_\rho$ (cf. [6; Theorem 8.5.7]), whence M_ρ is a proper, two-sided ideal. Hence $M_\rho \subseteq F$, and $\Sigma_{j=1}^{\infty} E_jFE_j \in F$. □

Using Proposition D, we can extend Corollary B to infinite, countably decomposable, semi-finite factors and prove it in sharpened form.

Corollary E.

Let M be an infinite, countably decomposable, semi-finite factor, $\mathcal{D}$ a numerical dimension function on the set of projections in M, $\{1',2',\ldots\}$ an infinite sequence of positive integers (not necessarily distinct), and H a self-adjoint operator in M. Then there is an orthogonal family $\{E_n\}$ of projections in M with sum I such that $n' \le \mathcal{D}(E_n) < n'+1$ and an orthogonal family $\{E_{1n},E_{2n},\ldots,E_{n'n}\}$ of equivalent projections in M with sum E_n such that $E_nHE_n = \Sigma_{j=1}^{n'} E_{jn}HE_{jn}$, and $\Sigma_{n\neq m} E_nHE_m$ is in the unique, proper, norm-closed ideal F of M.

Proof.

With F a projection of dimension 1, there are projections $G_1,G_2,\ldots$ as described in the statement of Theorem A. Since $\Sigma\mathcal{D}(G_j) = \mathcal{D}(\Sigma G_j) = \mathcal{D}(I) = \infty$ and $\mathcal{D}(G_j) \le 1$, we can find an orthogonal family of projections $\{E_n\}$, with sum I, such that $n' \le \mathcal{D}(E_n) < n'+1$, for some positive integer n', and each E_n is the sum of a subset of $\{G_1,G_2,\ldots\}$. By choice of $\{G_j\}$, $\Sigma_{j\neq k} G_jHG_k \in F$. From Proposition D, $\Sigma_{h=1}^{\infty} E_h(\Sigma_{j\neq k} G_jHG_k)E_h \in F$. Since each E_n is a sum of certain G_j,

$$\sum_{j\neq k} G_jHG_k - \sum_{h=1}^{\infty} E_h\left(\sum_{j\neq k} G_jHG_k\right)E_h = \sum_{n\neq m} E_nHE_m \in F .$$

As $n' \le \mathcal{D}(E_n) < n'+1$, we can find an orthogonal family $\{F_{1n}, F_{2n}, \ldots, F_{n'n}\}$ of equivalent projections in M with sum E_n. Using [5], as in the proof of Theorem A, we construct a unitary operator U in M that commutes with each E_n and satisfies

$$E_n H E_n = \sum_{j=1}^{n'} U^* F_{jn} U E_n H E_n U^* F_{jn} U .$$

We complete the proof by choosing $U^* F_{jn} U$ as E_{jn}. □

If we take $\mathcal{B}(H)$ as M, in the preceding corollary, we conclude that, for each bounded, self-adjoint operator, we can find an orthonormal basis relative to which the matrix of that operator has an arbitrarily preassigned system of diagonal blocks of finite size in diagonal form and the matrix, with its diagonal entries replaced by zero, is compact.

Remark F.

If the self-adjoint operator H on the separable Hilbert space H is represented as a matrix whose associated off-diagonal matrix is compact, then the essential spectrum and the essential norm of H (that is, the spectrum and norm of the image of H in the Calkin algebra) can be read from the diagonal of this matrix. Let $\mathbb{F}$ be the set of diagonal entries that appear at a finite number of diagonal positions, $\mathbb{S}$ be the closure of the set of all diagonal entries, and $sp_e(H)$ be $\mathbb{S} \setminus \mathbb{F}$. Then $sp_e(H)$ is the essential spectrum of H and $\sup\{|\lambda| : \lambda \in sp_e(H)\}$ $(= ||H||_e)$ is the essential norm of H. If $||H||_e = 0$, then H is compact. In any case, $||H||_e$ is the (minimum) distance from H to the ideal of compact operators on H. If $||H||_e \neq 0$, the special matrix representation described provides a ready means for constructing (many) compact approximants to H realizing this distance. For general bounded operators on H, the polar decomposition can be used, in conjunction with the preceding construction, to produce compact approximants. Similar comments apply to countably decomposable II_∞ factors and their unique, proper, norm-closed ideals. □

REFERENCES

1. Berg, I.D. (1971). An extension of the Weyl-von Neumann theorem to normal operators. Trans. Amer. Math. Soc., 160, pp. 365-371.
2. Halmos, P.R. (1970). Ten problems in Hilbert space. Bull. Amer. Math. Soc., 76, pp. 887-933.
3. Halmos, P.R. (1972). Continuous functions of Hermitian operators. Proc. Amer. Math. Soc., 31, pp. 130-132.

4. Halmos, P.R. (1968). Quasitriangular operators. Acta Sci. Math. (Szeged), 29, pp. 285-294.
5. Kadison, R.V. (1984). Diagonalizing matrices, Amer. J. Math., 106, pp. 1451-1468.
6. Kadison, R.V. & Ringrose, J.R. (1983 and 1986). Fundamentals of the Theory of Operator Algebras, Vols. I and II. Academic Press, New York.
7. Kaftal, V. (1978). On the theory of compact operators in von Neumann algebras II. Pac. J. Math., 79, pp. 129-137.
8. Neumann, J. von. (1935). Charakterisierung des Spektrums eines Integraloperators. Actualités Sci. Indust., no. 229, Hermann, Paris.
9. Weyl, H. (1909). Über beschränkte quadratische Formen, deren Differenz vollstetig ist, Rendiconti del circolo Matematico di Palermo, 27, pp. 373-392.
10. Zsido, L. (1975). The Weyl-von Neumann theorem in semifinite factors. J. Func. Analysis, 18, pp. 60-72.

SECONDARY INVARIANTS FOR ELLIPTIC OPERATORS AND OPERATOR ALGEBRAS

by

JEROME KAMINKER [1]

§1 Introduction

If D is an elliptic operator on a closed manifold then its index is an invariant which depends only on the principal symbol. We will discuss here a means of producing secondary invariants which depend on more than the principal symbol. They will be generalizations of the relative η-invariant of Atiyah-Patodi-Singer. That invariant corresponds to the transgressed Chern character. The ones that we define will also correspond to certain secondary characteristic classes. They will be invariants to detect a certain class of perturbations of the operator. Operator algebras come into the picture via the choice of the class of perturbations. Roughly, we will embed the elliptic operator D into a "family" so that the index as an element of the K-theory of an algebra associated to that family. The invariants will be obtained from that K-theory element.

This is an application of work with Ron Douglas and Steve Hurder. It is related to the longitudinal part of our theory. The ideas here are also heavily influenced, both implicitly and explicitly, by Alain Connes' view of the subject.

§2 Higher order invariants of elliptic operators

In this section we will discuss how one might view secondary invariants for operators. Suppose that D and D' are k^{th} order elliptic differential operators acting on the same vector bundles, $D,D':C^\infty(E) \longrightarrow C^\infty(F)$. The order of D-D' is less than k if and only if D and D' have the same principal symbol. In this case one has Index(D) = Index(D'). A higher order invariant will be one which is able to distinguish between D and D' even if they have the same principal symbol.

If we fix D, then D' is a perturbation of D. Certain types of

[1]Research supported in part by a grant from NSF.

perturbations will lead to computable invariants. We take D' to be D with coefficients in a trivialized flat bundle. Specifically, let E be a Hermitian bundle and $D:C^\infty(E) \longrightarrow C^\infty(E)$ a self-adjoint elliptic operator. Let $\alpha:\pi_1(M) \longrightarrow U_n$ be a representation of the fundamental group of M and let $F_\alpha = \tilde{M}x_\alpha \mathbb{C}^n$ be the flat vector bundle determined by α with connection ∇_α. Then D extends to an operator $D\otimes_{\nabla_\alpha} I:C^\infty(E\otimes F_\alpha) \longrightarrow C^\infty(E\otimes F_\alpha)$. Let $\Theta:\tilde{M}x_\alpha \mathbb{C}^n \longrightarrow Mx\mathbb{C}^n$ be a trivialization. Then θ induces $\theta^*:C^\infty(E\otimes F_\alpha) \longrightarrow C^\infty(E\otimes(Mx\mathbb{C}^n))$. We view the operator $\theta^*(D\otimes_{\nabla_\alpha} I)(\theta^*)^{-1} = D_\alpha$ as a perturbation of $D\otimes I_n$ since they act on the same space. In fact, $D_\alpha - D\otimes I_n$ is a 0th-order operator given by a section Φ of $End(E\otimes(Mx\mathbb{C}^n))$ of the form $I\otimes\varphi$.

Now, fix α and θ and choose $\rho \in \hat{U}_n$, $\rho:U_n \longrightarrow U_N$, where $\hat{U}_n$ denotes the irreducible unitary representations of U_n. Then $\{D_{\rho\alpha}\}$ is a family of operators parameterized by $\hat{U}_n$. Each $D_{\rho\alpha}$ has the same principal symbol as $D\otimes I_N$ where $N = \deg\rho$. We shall obtain a family of invariants which will distinguish $D_{\rho\alpha}$ from $D\otimes I_N$.

§3 Lifting the operator to a foliation

The next step is to associate to D an operator on a foliated manifold. The pair (α,Θ) determined a flat principal U_n bundle with trivialization $\tilde{M}x_\alpha U_n \cong MxU_n$. It has the foliation $\mathcal{F}_\alpha$ with leaves the images of $\tilde{M}x\{g\}$. The operator $D:C^\infty(E) \longrightarrow C^\infty(E)$ can be lifted to an operator $D_\alpha:C^\infty(\pi^*(E)) \longrightarrow C^\infty(\pi^*(E))$, where $\pi:MxU_n \longrightarrow M$. It is leafwise elliptic with respect to $\mathcal{F}_\alpha$ and self-adjoint. Hence, it defines an element $[D_\alpha] \in KK^1(C(V),C^*(V,\mathcal{F}_\alpha))$, [7], [9]. The element $[D_\alpha]$ depends on the principal symbol of D and the data (α,θ). It can be thought of as D "perturbed" by (α,θ). This is accurate in the following sense. If D_α is viewed as an unbounded operator on $L^2(\pi^*(E)x_\alpha U_n) \cong L^2(ExU_n) \cong \sum_\rho L^2(E)\otimes V_\rho$ then it decomposes as $\sum_\rho \theta^*(D\otimes_{\nabla_{\rho\alpha}} I)(\theta^*)^{-1}$, where

$\theta^*:L^2(E\times U_n) \longrightarrow L^2(\pi^*(E)\times_\alpha U_n)$. Note that $\theta^*(D\otimes_{\nabla_{\rho\alpha}} I)(\theta^*)^{-1} = D\otimes I + \Phi_{\theta,\alpha,\rho}$ where $\Phi_{\theta,\alpha,\rho}$ is a 0th-order operator. Thus, D_α corresponds to the entire family of perturbations of D parameterized by $\hat{U}_n$.

Next, we pair $[D_\alpha]$ with an element $[u] \in KK^1(\mathbb{C},C(M\times U_n))$ to obtain $[u]\otimes_{C(M\times U_n)}[D_\alpha] \in KK(\mathbb{C},C^*(M\times U_n,\mathcal{F}_\alpha))$. We will be particularly interested in elements [u] of the form $pr_2^*([\rho])$ where $\rho:U_n \longrightarrow U_N$ is a representation of U_n, and $pr_2:M\times U_n \longrightarrow U_n$. Now, $C^*(M\times U_n,\mathcal{F}_\alpha) \cong (C(U_n)\rtimes\Gamma)\otimes\mathcal{K}$ where Γ denotes the image of $\alpha:\pi_1(M) \longrightarrow U_n$, so Haar measure, μ, on U_n will induce a trace on $C^*(M\times U_n,\mathcal{F}_\alpha)$ which we will call Tr_μ. The next task is to interpret $Tr_\mu([u]\otimes_{C(M\times U_n)}[D_\alpha])$.

§4 Toeplitz index theory along the leaves of a foliation

On a closed manifold there is a way to associate an algebra of Toeplitz operators to an elliptic self-adjoint operator D, [3]. Letting P denote the orthogonal projection onto the non-negative eigenspaces of D, the Toeplitz algebra is $\mathcal{T} = \{T_\varphi = PM_\varphi P \mid \varphi \in C(M)\}$ acting on $L^2(M)$. There is an index formula which depends on the symbol of D and the multiplier φ. On an open manifold the same algebra could be constructed, but φ being invertible would no longer yield that T_φ is Fredholm. Suppose, however, that the manifold is a leaf in a foliation with an invariant transverse measure. Then φ invertible implies that T_φ is invertible modulo the von Neumann algebra of the foliation and the (real-valued) index of T_φ is given by a topological formula.

A precise formulation goes as follows. Let V be the manifold foliated by $\mathcal{F}$ and let D be the leafwise elliptic self-adjoint differential operator. Identifying $KK^1(C(V),C^*(V,\mathcal{F})) \cong Ext(C(V),C^*(V,\mathcal{F}))$ one observes that [D] corresponds to the extension $[\mathcal{T}_\epsilon]$,

$$0 \longrightarrow C^*(V,\mathcal{F}_\alpha) \longrightarrow \mathcal{T}_\epsilon \longrightarrow C(V) \longrightarrow 0$$

where $\mathcal{T}_\epsilon$ is the C*-algebra generated by operators of the form $P_\epsilon M_\varphi P_\epsilon$.

Here, $P_\epsilon = h_\epsilon(D)$ is a smoothed version of the positive projection, i.e. the function h_ϵ is smooth, increasing, and agrees with $\chi_{[0,\infty)}$ outside of $(-\epsilon,\epsilon)$. On the other hand, the genuine Toeplitz extension along the leaves is

$$0 \longrightarrow \mathcal{C} \longrightarrow \mathcal{T} \longrightarrow C(V) \longrightarrow 0$$

where $\mathcal{C}$ is the commutator ideal of $\mathcal{T}$. Then φ invertible implies that the projections onto the kernel of T_φ and the kernel of T^*_φ are in the finite trace ideal and hence $\operatorname{Index}_\mu(T_\varphi) = \operatorname{Tr}_\mu(P_{\ker T_\varphi}) - \operatorname{Tr}_\mu(P_{\ker T_\varphi{}^*})$ is defined.

<u>Theorem 1</u>: [9] There is a cyclic cocycle on the algebra $C^\infty(V)$, defined by $C^L_D(\varphi_0,\ldots,\varphi_p) = \operatorname{Tr}_\mu(P[P,\varphi_0]\ldots[P,\varphi_p])$ and satisfying $\operatorname{Index}_\mu(T_\varphi) = \langle[C^L_D],[\varphi]\rangle$.

Before stating the topological formula for the index we recall the ingredients. If $D:C^\infty(E) \longrightarrow C^\infty(E)$ is a self-adjoint leafwise elliptic operator then the symbol yields a map $\sigma_1(D):S\mathcal{F} \longrightarrow U_N$, where $S\mathcal{F}$ is the unit sphere bundle of $T\mathcal{F}$, and hence an element $[\sigma_1(D)] \in K_1(C(S\mathcal{F}))$. The cohomology algebra $H^*(U_N,\mathbb{C}) \cong \Lambda(y_1,\ldots,y_{2N-1})$, where y_i is represented by the differential form $(\text{const.})\operatorname{Tr}((\Theta^{-1}d\Theta)^{2i+1})$, $\Theta^{-1}d\Theta$ the Maurer-Cartan form for U_N. Then $\operatorname{ch}([\sigma_1(D)])$ is represented by $\sigma_1(D)^*(1+y_1+\ldots)$, a sum of differential forms on $S\mathcal{F}$. Similarly, if $[\varphi] \in K_1(C(V))$, then $\operatorname{ch}([\varphi]) = \varphi^*(1+y_1+\ldots)$. Finally, using the metric on $T\mathcal{F}\otimes\mathbb{C}$ chosen initially we obtain a connection and represent the Pontrijagin classes, and hence $\operatorname{Todd}(T\mathcal{F}\otimes\mathbb{C})$ by differential forms on V. Let $\pi:S\mathcal{F} \longrightarrow V$ and consider the form $\operatorname{ch}(\sigma_1(D))\wedge\pi^*(\operatorname{Todd}(T\mathcal{F}\otimes\mathbb{C})\wedge\operatorname{ch}(\varphi))$ on $S\mathcal{F}$. Let $\langle\operatorname{ch}(\sigma_1(D))\wedge\pi^*(\operatorname{Todd}(T\mathcal{F}\otimes\mathbb{C})\wedge\operatorname{ch}(\varphi)),[C_\mu]\rangle$ denote its evaluation against the Ruelle-Sullivan current associated to μ lifted to $S\mathcal{F}$. One then has

<u>Theorem 2</u>: [9] $\operatorname{Index}_\mu(T_\varphi) = \langle\operatorname{ch}(\sigma_1(D))\wedge\pi^*(\operatorname{Todd}(T\mathcal{F}\otimes\mathbb{C})\wedge\operatorname{ch}(\varphi)),[C_\mu]\rangle$

Now it follows from the index theorem of Connes and Skandalis, [7], that $\mathrm{Index}_\mu(T_\varphi) = \langle[\mathrm{Tr}_\mu],[\varphi]\otimes_{C(V)}[D]\rangle$, (c.f. Remark 4.15). We will apply these relations in the next section, but it is interesting to note here that from Theorem 1 and the above remark one has $\langle[C_D^L],[\varphi]\rangle = \langle[\mathrm{Tr}_\mu],[\varphi]\otimes_{C(V)}[D]\rangle$. This suggests on a formal level that there ought to be a bivariant cyclic theory, a bivariant Chern character, and a Kasparov like pairing which would yield $[C_D^L] = [\mathrm{Tr}_\mu]\#\mathrm{Ch}([D])$, (c.f. [8]). One could then obtain $\langle[\mathrm{Tr}_\mu],[\varphi]\otimes_{C(V)}[D]\rangle = \langle[\mathrm{Tr}_\mu]\#\mathrm{Ch}([D]),[\varphi]\rangle$. It thus becomes natural to pair the character of the operator with cyclic classes other than the trace.

§5 Secondary invariants

We can now define secondary invariants for the operator D. The setup is again as in §3. Given (α,θ) we obtain $[D_\alpha] \in KK^1(C(M\mathrm{x}U_n),C^*(M\mathrm{x}U_n,\mathcal{F}_\alpha))$. Let $[\mathcal{Z}] \in HC^*(C_c^\infty(\mathcal{F}_\alpha))$, where $C_c^\infty(\mathcal{F}_\alpha)$ is the smooth convolution algebra of the foliation. Assume that it pairs with $K_*(C^*(M\mathrm{x}U_n,\mathcal{F}_\alpha))$. The secondary invariant for D associated to (α,Θ) and $[\mathcal{Z}]$ is a functional on $K_1(C(M\mathrm{x}U_n))$,

$$\mathcal{Z}(D,\alpha,\theta):K_1(C(M\mathrm{x}U_n)) \longrightarrow \mathbb{C}$$

defined by $\mathcal{Z}(D,\alpha,\theta)([\varphi]) = \langle[\mathcal{Z}],[\varphi]\otimes[D_\alpha]\rangle$. We will justify this terminology by showing that for a certain choice of $[\mathcal{Z}]$ one gets a known secondary invariant.

Recall the η-invariant of an elliptic self-adjoint operator D, [1]. It is defined to be the value of the meromorphic extension of $\eta(D,s) = \sum_{\lambda_i\neq 0} \frac{\mathrm{sign}(\lambda_i)}{(\lambda_i)^{-s}}$ at $s = 0$, where λ_i are the eigenvalues of D. It is denoted by $\eta(D)$. Given (α,θ), and a representation ρ, the relative η-invariant is $\eta(D,\rho,\alpha,\theta) = \eta(D\otimes_{\nabla_\alpha}I) - \eta(D\otimes I)$. Although the operators $D\otimes_{\nabla_\alpha}I$ and $D\otimes I$ have the same principal symbol, they can be distinguished by the relative η-invariant.

A principal result of [10] implies that the relative η-invariant is a secondary invariant in the sense described above. Let $[\varphi] = \mathrm{pr}_2^*([\rho])$, $[\rho]$

$\in K^1(U_n)$. Then $[\varphi]\otimes_{C(M\times U_n)}[D_\alpha] \in K_0(C^*(M\times U_n,\mathcal{F}_\alpha))$.

Theorem 4: Consider $[Tr_\mu] \in HC^0(C_c^\infty(\mathcal{F}_\alpha))$. Then one has

$$<[Tr_\mu],pr_2^*[\rho]\otimes_{C(M\times U_n)}[D_\alpha]> = \eta(D,\rho,\alpha,\theta).$$

Proof: It follows from [7], that $<[Tr_\mu],pr_2^*[\rho]\otimes_{C(M\times U_n)}[D_\alpha]> = \mathrm{Index}(T_\varphi)$, where $\varphi = \rho pr_2$. By Theorem 2, [9], one sees that $\mathrm{Index}(T_\varphi) = <[C_D^L],[\varphi]>$. Finally, it is shown in [10] that $<[C_D^L],[\varphi]> = \eta(D,\rho,\alpha,\theta)$. ∎

The results of [10] provide a new proof of the Index Theorem for Flat Bundles [1], which states that $\eta(D,\rho,\alpha,\theta) = \int_{S(M)} ch(\sigma_1(D))\wedge\pi^*(Todd(TM\otimes\mathbb{C})\wedge Tch(\rho,\alpha,\Theta)))dm$. One can show directly that this is equal to $<ch(\sigma_1(D))\wedge\pi^*(Todd(T\mathcal{F}\otimes\mathbb{C})\wedge ch(\varphi)),[C_\mu]> = <[Tr_\mu],pr_2^*[\rho]\otimes_{C(M\times U_n)}[D_\alpha]>$. If one uses this, then Theorem 4 follows directly from [9], avoiding the more complicated analysis in [10].

It is worth noting that a family of cyclic classes parameterized by Gelfand-Fuchs cohomology has been constructed by Connes [6]. Thus, our invariants are obtained by lifting the operator to a foliation and then pairing with cyclic classes which include among them those corresponding to secondary characteristic classes for foliations.

More precisely, in [6], Connes constructs a map $c:H^*(WO_m) \longrightarrow HC^*(C^\infty(U_n)\rtimes\Gamma)$, $m=\dim U_n$, where Γ is the image of $\alpha:\pi_1(M) \longrightarrow U_n$. Now, $C^*(M\times U_n,\mathcal{F}_\alpha) \cong (C(U_n)\rtimes\Gamma)\otimes\mathcal{K}$ and for the appropriate choice of dense subalgebra $C^\infty(\mathcal{F}_\alpha)$ (see e.g. [11]) there is a map $c:H^*(WO_m) \longrightarrow HC^*(C^\infty(\mathcal{F}_\alpha))$. Connes has shown that the image of c pairs with $K_*(C^*(M\times U_n,\mathcal{F}_\alpha))$. Thus for any $\chi \in H^*(WO_m)$ we obtain a secondary invariant $\chi(D,\alpha,\Theta)$,

$$\chi(D,\alpha,\theta)([\varphi]) = <c(\chi),[\varphi]\otimes_{C(M\times U_n)}[D_\alpha]>.$$

Note that the relative η-invariant is not obtained from the image of c since it corresponds to the transgressed Chern character, Tch, which is an element of the complex WO_m, but is not a cycle there, (thus does not define an element of $H^*(WO_m)$). It does however define a leaf class (c.f. [11]) and this can be paired with the index class which can be viewed as living on a transversal.

The secondary invariants can be viewed as elements of $H_*(MxU_n)$ and their components in the Kunneth decomposition can be analyzed. This is done in more detail in [10].

REFERENCES

1. M. F. Atiyah, V. K. Patodi, and I. M. Singer, "*Spectral asymmetry and Riemannian geometry I, II, III*", Math. Proc. Camb. Phil. Soc. 77(1975), 43-69, 78(1975), 405-432, 79(1976), 71-99

2. P. Baum and A. Connes, "*Leafwise homotopy equivalence and rational Pontryjagin classes*", Foliations, Adv. Studies in Pure Mathematics 5, 1985,1-14

3. P. Baum and R. G. Douglas, "*K homology and index theory*", Proc. Symposia Pure Math. **38** (1982), Part I, American Math. Society, Providence, 117-174.

4. A. Connes, "*Non-commutative differential geometry-I: the Chern character in K-homology*", Publ. Math. I.H.E.S. **62** (1986), 41-93

5. ________ , "*Non-commutative differential geometry-II: De Rham homology and non-commutative algebra*", Publ. Math. I.H.E.S. **62** (1986), 94-144.

6. ________ , "*Cyclic cohomology and the transverse fundamental class of a foliation*", 52-144, Geometric Methods in Operator Algebras, Ed. by H. Araki and E. G. Effros, Longman Scientific & Technical, Great Britain, 1986

7. A. Connes and G. Skandalis, "*The longitudinal index theorem for foliations*", Publ. RIMS, Kyoto Univ. **20** (1984), 1139-1183

8. R. G. Douglas, "*Elliptic invariants and operator algebras: toroidal examples*", this volume

9. R. G. Douglas, S. Hurder, and J. Kaminker,"*The longitudinal cyclic cocycle and the index of Toeplitz operators*", in preparation

10. ____________________________,"*The η-invariant and the transverse Chern character*", in preparation

11. G. Elliott, T. Natsume, and R. Nest, "*Cyclic cohomology for one-parameter smooth crossed products*", pre-print, 1987

12. S. Hurder, "*Global invariants for measured foliations*", Transactions Amer. Math. Soc., 280 (1983), 367-391

13. J. Kaminker and J. G. Miller, "*Homotopy invariance of the analytic index of signature operators over C^*-algebras*", J. Operator Theory, 14(1985), 113-127

Address of author:

Department of Mathematical Sciences
IUPUI
Indianapolis, IN 46223
USA

INVERSE LIMITS OF C^*-ALGEBRAS AND APPLICATIONS

N. Christopher Phillips[1]

University of California at Los Angeles, Los Angeles, CA 90024

Introduction. The purpose of this article is to advertise inverse limits of C^*-algebras. These are complete topological $*$-algebras whose topology is determined by their continuous C^*-seminorms. Recently such algebras have appeared in a number of problems whose formulation involves only C^*-algebras. As examples we mention the study of the multipliers of the Pedersen ideal of a C^*-algebra, the construction of "noncommutative" analogs of noncompact classical Lie groups, the construction of the "noncommutative loop space" of a C^*-algebra, and the comparison of the K-theory of crossed products by homotopic actions of compact Lie groups. In some of these, particularly the last one, it is not at all evident that inverse limits of C^*-algebras should appear.

Inverse limits of Banach algebras have been studied sporadically since 1952, when they were introduced by Arens [6] and Michael [46]. Inverse limits of C^*-algebras first appeared in [6], and the first major theorems concerning them appear in papers of C. Wenjen [73] and Xia Daoxing [75]. They have since been studied, under various names, by Brooks [16], Inoue [36], Schmüdgen [63], Fritsche [34], [35], Mallios [44], [45], and Fragoulopoulou [30], [31], [32]. It should be noted that some of these authors used equivalent definitions but did not use inverse limits. Also, the algebras to which the main theorems of [44] and [45] apply are required to satisfy an additional condition, which in [53], Proposition 1.14, is shown to imply that they are actually C^*-algebras. There has furthermore been work on larger

[1]Research partially supported by a National Science Foundation postdoctoral fellowship.

classes of topological *-algebras, which satisfy some analog of the C^* property but need not have continuous C^*-seminorms, by Allan [4], Dixon [26], and others.

It was, however, Arveson in [7] who, without being aware of previous work on inverse limits of C^*-algebras, gave the first interesting examples of such algebras which were relevant to the study of C^*-algebras, namely his tangent algebras. (Actually, the multiplier algebra of the Pedersen ideal had been studied earlier, but had not been recognized as an inverse limit of C^*-algebras.) Further examples of inverse limits of C^*-algebras relevant to the study of C^*-algebras, namely the noncommutative analogs of the algebras of continuous functions on certain classical Lie groups, were given by Voiculescu in [71]. In [55], we used inverse limits of C^*-algebras to establish a theorem on the invariance, under homotopy of the action, of the completed K-theory of a crossed product by a compact Lie group. In [56], we showed that the multiplier algebra of the Pedersen ideal, which had previously been studied in [41], is an inverse limit of C^*-algebras, and we used this result to simplify a number of the proofs in [41].

A number of terms have been used for various kinds of inverse limits of C^*-algebras, including b^*-algebra, F^*-algebra, locally C^*-algebra, and LMC^*-algebra. Here we follow Voiculescu and call the objects we study pro-C^*-algebras. (Strictly speaking, Voiculescu's pro-C^*-algebras are equivalent to, but not identical to, our inverse limits of C^*-algebras. Following Voiculescu's suggestion, however, we call our objects pro-C^*-algebras anyway. It should also be pointed out that a pro-C^*-algebra is not the same thing as a pro-object in the category of C^*-algebras, as defined in [9].) We also adopt Arveson's term σ-C^*-algebra for countable inverse limits of C^*-algebras.

This article is divided into two parts. In the first part, we describe some of the general theory of pro-C^*-algebras. In the first section of this part, we give the basic definitions and some examples. In the second section we discuss more of the basic theory, namely the spectrum, functional calculus, and the algebra of bounded elements. The third section is devoted to the universal pro-C^*-algebra on a set of generators and relations; the importance of this notion stems from the fact that many sets of generators and relations which do not satisfy the conditions of [11], and hence do not admit universal C^*-algebras, nevertheless do admit universal pro-C^*-algebras. In the fourth section we describe the commutative unital pro-C^*-algebras, and the fifth section contains the analogs for pro-C^*-algebras of such things as tensor products, free products, and multiplier algebras. The last section is devoted to the appropriate version of K-theory for σ-C^*-algebras.

The second part of this article contains the applications. We discuss, in order and with varying amounts of detail, Arveson's tangent algebra [7], the multipliers of Pedersen's ideal [41], [56], K-theory of crossed products by homotopic actions of compact Lie groups [55], "noncommutative" classical Lie groups [71], classifying spaces for the K-theory of C^*-algebras, and the "noncommutative loop space". The last two items appear here for the first time (except for one remark about classifying spaces in [54]), and we are able to generalize and simplify the construction of the tangent algebra; for the other three applications, we merely give surveys.

For many results we give only sketches of proofs, or no proofs at all, especially toward the end of the first part and in the more complicated applications. For omitted proofs in the general theory of pro-C^*-algebras, we refer to [53]; we have included here proofs of most of the results quoted from the earlier literature in that paper. The results on K-theory in the last section of Part 1 are all proved in [54], and, for the applications, complete proofs can be found in the papers where they first appear.

We should at least mention two topics that are not discussed here, namely Hilbert modules and representations. The theory of Hilbert modules over pro-C^*-algebras is developed in [53], and they play a role in the K-theory of σ-C^*-algebras. Representations of pro-C^*-algebras are studied in [16], [36], and [32], but play no role in any of the applications considered here. Also, there are several applications which we do not discuss. One, involving Spanier-Whitehead duality and locally trivial continuous trace algebras over finite complexes, is contained in work in progress by Ellen Parker and Bob Rigdon. Another, involving KK-theory and index theory, appears in the thesis of Jens Weidner [72]. Finally, it is possible that the σ-C^*-algebra $q^\infty A = \varprojlim q^n A$ (compare [21]) could have applications to K-theory.

Part 1: Pro-C^*-Algebras.

1.1. Definitions and examples. An inverse system (also called a projective system) of sets consists of a directed set D, a set X_d for each $d \in D$, and functions $\pi_{d,e} : X_d \to X_e$ for all $d, e \in D$ with $d \geq e$, satisfying the consistency conditions $\pi_{d,d} = \mathrm{id}_{X_d}$ and $\pi_{e,f} \circ \pi_{d,e} = \pi_{d,f}$ for $d \geq e \geq f$. The inverse limit (or projective limit) of the system (X_d) is a set X together with functions $\kappa_d : X \to X_d$ such that $\pi_{d,e} \circ \kappa_d = \kappa_e$ for $d \geq e$, and satisfying the following universal property: given any set Y and functions $\varphi_d : Y \to X_d$ such that $\pi_{d,e} \circ \varphi_d = \varphi_e$, there is a unique function $\varphi : Y \to X$ such that $\varphi_d = \kappa_d \circ \varphi$ for all d. The inverse limit X is denoted by $\varprojlim X_d$, and it is clearly unique up to unique isomorphism. That it exists is seen by taking

$$(*) \qquad X = \{(x_d) \in \prod_{d \in D} X_d : \pi_{d,e}(x_d) = x_e \text{ whenever } d \geq e.\},$$

with κ_d being the projection on the factor X_d. This observation enables one to identify elements of the inverse limit $\varprojlim X_d$ with the "coherent sequences" $(x_d)_{d\in D}$ satisfying $\pi_{d,e}(x_d) = x_e$ for $d \geq e$.

The definition of an inverse limit generalizes in an obvious way to any category, and it is easily seen that the same construction produces an inverse limit in the categories of abelian groups and homomorphisms, topological spaces and continuous maps, and topological *-algebras over $\mathbf{C}$ and continuous *-homomorphisms. (The product in (*) is given the product topology and the pointwise algebraic operations.)

1.1.1. PROPOSITION. ([63]) *Let A be a topological *-algebra over* **C**. *Then A is isomorphic, as a topological *-algebra, to an inverse limit of C^*-algebras in the above sense if and only if A is Hausdorff, complete, and its topology is determined by the set of all continuous C^*-seminorms on A.*

A C^*-seminorm on A is of course a seminorm p satisfying $p(ab) \leq p(a)p(b)$, $p(a^*) = p(a)$, and $p(a^*a) = p(a)^2$ for all $a, b \in A$. The topology determined by a set S of seminorms on A is the topology in which a net (a_α) converges to 0 if and only if $p(a_\alpha) \to 0$ for all $p \in S$. We also note that the inverse limit above is taken in the category of topological *-algebras; inverse limits do turn out to exist in the category of C^*-algebras, but the inverse limit in the category of C^*-algebras is not the same as the inverse limit in the category of topological *-algebras, and in particular is not given by $(*)$.

Proof of Proposition 1.1.1. Let A be complete and Hausdorff, and let its topology be determined by the set S of all continuous C^*-seminorms on A. For $p \in S$, let A_p be the Hausdorff completion of A in the seminorm p. Then A_p is a C^*-algebra, and there is an obvious homomorphism $\varphi : A \to \prod_{p \in S} A_p$ whose image is obviously contained in the inverse limit $\varprojlim_{p \in S} A_p$ (with respect to the obvious order on S, namely $p \leq q$ if $p(a) \leq q(a)$ for all a, and the obvious maps $A_q \to A_p$). The map φ is injective, because A is Hausdorff, and is a homeomorphism onto its image, because the topology on A is determined by S. It remains to show that $\varphi(A) = \varprojlim_{p \in S} A_p$. Let $(a_p)_{a \in S}$ be a coherent sequence, and, for $p \in S$ and $n \in$ **N**, choose $x_{p,n} \in A$ whose image in A_p differs from a_p by less than $\frac{1}{n}$ in norm. Then the net $(x_{p,n})$ is Cauchy in A, and hence has a limit x, whose image in $\prod A_p$ is (a_p).

For the converse, let $A = \varprojlim A_d$ be an inverse limit of C^*-algebras, with canonical maps $\kappa_d : A \to A_d$. Let p_d be the continuous C^*-seminorm on A given by $p_d(a) = \|\kappa_d(a)\|$. If p is any continuous seminorm on A, then it is easily seen that there is $c > 0$ and $d_1, \ldots, d_n \in D$ such that $p(a) \leq c \max(p_{d_1}(a), \ldots, p_{d_n}(a))$ for $a \in A$. If p is a C^*-seminorm, then it follows that $p \leq p_d$ when $d \in D$ is chosen such that $d \geq d_1, \ldots, d_n$. It follows that the topology on A is determined by all of its continuous C^*-seminorms. Furthermore, A is complete and Hausdorff because it is a closed subset of the complete Hausdorff topological *-algebra $\prod_{d \in D} A_d$. Q.E.D.

1.1.2. DEFINITION ([53], [7]). A topological *-algebra satisfying the conditions of the previous proposition is called a ***pro-C^*-algebra.*** If it is a countable inverse limit of C^*-algebras (equivalently, if its topology is determined by countably many continuous C^*-seminorms), then it is called a ***σ-C^*-algebra.***

Since an inverse limit is unchanged if the inverse system is restricted to a cofinal subset, and since every countable directed set with no largest element has a cofinal subset isomorphic to $\mathbf{N}$, we see that all σ-C^*-algebras can be expressed as inverse limits $\varprojlim_{n \in \mathbf{N}} A_n$. Note that the maps $\pi_{m,n} : A_m \to A_n$ are determined by the maps $\pi_n : A_{n+1} \to A_n$, and that the maps π_n can be arbitrary.

As will be seen on several occasions below, σ-C^*-algebras are in general much better behaved then pro-C^*-algebras. One major reason for this is the possibility of constructing coherent sequences by induction over the index set. For example, if $(X_n)_{n \geq 0}$ is an inverse system of sets in which each map $X_{n+1} \longrightarrow X_n$ is surjective, then each map $\varprojlim X_n \longrightarrow X_n$ is surjective. However, there exists an inverse system (X_d) of sets in which each map $X_d \longrightarrow X_e$ is surjective, each X_d is nonempty, and yet $\varprojlim X_d = \emptyset$. (Also see Example 1.4.8.)

1.1.3. *Notation.* If A is a pro-C^*-algebra, then $S(A)$ is the set of all continuous C^*-seminorms on A, ordered as in the proof of Proposition 1.1.1. For $p \in S(A)$, we let $\mathrm{Ker}(p) = \{a \in A : p(a) = 0\}$ and we let A_p be the completion of $A/\mathrm{Ker}(p)$ in the norm determined by p.

Note that A_p is a C^*-algebra, and that the proof of Proposition 1.1.1 gives a canonical representation of A as an inverse limit, namely $A \cong \varprojlim_{p \in S(A)} A_p$. ($S(A)$ is directed because if $p, q \in S(A)$, then $a \mapsto \max(p(a),\ q(a))$ is also in $S(A)$.) We will show in Corollary 1.2.8 that in fact $A/\mathrm{Ker}(p)$ is always already complete.

1.1.4. EXAMPLES. (1) Every C^*-algebra is a pro-C^*-algebra.

(2) A closed *-subalgebra of a pro-C^*-algebra is a pro-C^*-algebra.

(3) The product $\prod_{\alpha\in I} A_\alpha$ of C^*-algebras, with the product topology, is a pro-C^*-algebra. (The elements of the product are *all* families $(a_\alpha)_{\alpha\in I}$ with $a_\alpha \in A_\alpha$, not just the bounded families. The continuous C^*-seminorms $p_\beta((a_\alpha)_{\alpha\in I}) = \|a_\beta\|$ determine the topology.)

(4) Let X be a compactly generated Hausdorff space ([74], Section I.4). This means that a subset $C \subset X$ is closed if and only if $C \cap K$ is closed for every compact subset K of X. Then the algebra $C(X)$ of all continuous (not necessarily bounded) complex-valued functions on X, with the topology of uniform convergence on compact subsets, is a pro-C^*-algebra. Note that, according to [74], I.4, all metrizable spaces and all locally compact Hausdorff spaces are compactly generated.

Another important general construction of pro-C^*-algebras is discussed in Section 1.3. Also, Part 2 of this article can be regarded as a list of examples of pro-C^*-algebras together with explanations of what they are good for.

The homomorphisms between pro-C^*-algebras are, of course, taken to be the continuous *-homomorphisms. The definition of an inverse limit shows that the homomorphisms from a pro-C^*-algebra A to some other pro-C^*-algebra are determined by a knowledge of the homomorphisms from A to C^*-algebras. These are described as follows:

1.1.5. LEMMA. *Let A be a pro-C^*-algebra, and let B be a C^*-algebra. Then $Hom(A,B) \cong \varinjlim_{p\in S(A)}$* $\mathrm{Hom}(A_p, B)$. *If $\kappa_p : A \to A_p$ is the canonical map, then the map from* $\mathrm{Hom}(A_p, B)$ *to* $\mathrm{Hom}(A, B)$ *is $\varphi \mapsto \varphi \circ \kappa_p$.*

The proof is straightforward, using the fact that if $\varphi : A \to B$ is a (continuous) homomorphism, then $a \mapsto \|\varphi(a)\|$ is a continuous C^*-seminorm on A.

Unfortunately, it is not in general true that *-homomorphisms are automatically continuous. See Example 1.4.9. However, for σ-C^*-algebras, we have:

1.1.6. THEOREM ([53]). *Let A be a σ-C^*-algebra and let B be a pro-C^*-algebra. Then every *-homomorphism from A to B is continuous.*

For the proof, note that, by the previous lemma, we may take B to be a C^*-algebra. Representing B faithfully on a Hilbert space H, we may take $B = L(H)$, the set of all bounded operators on H. Unitizing A (see Definition 1.2.1 below), we reduce to the case of Lemma 3.1 of [13]. The key step in the proof of that lemma follows from a theorem of Xia Daoxing [75], according to which every positive linear functional on a unital σ-C^*-algebra is continuous. A proof of this result in English appears in Section 11 of [24].

1.2. **Functional calculus and bounded elements.** To define functional calculus, we need the spectrum of an element, and to define the spectrum in a nonunital pro-C^*-algebra, we need the unitization.

1.2.1. DEFINITION ([36], [53]). Let A be a pro-C^*-algebra. Then the *unitization* A^+ of A is the *-algebra $A \oplus \mathbf{C}$ with multiplication given by $(a, \lambda)\cdot(b, \mu) = (ab + \lambda b + \mu a, \lambda\mu)$, adjoint operation $(a, \lambda)^* = (a^*, \bar{\lambda})$, and the usual topology and vector space operations.

Note that if $A = \varprojlim A_d$, then $A^+ = \varprojlim A_d^+$; it follows that the unitization of a pro-C^*-algebra is again a pro-C^*-algebra.

We can now define the spectrum $sp(a)$ of an element a of a pro-C^*-algebra A in exactly the same manner as for a C^*-algebra. However, $sp(a)$ need be neither closed nor bounded. In fact, if $S \subset \mathbf{C}$ is any nonempty subset whatsoever, then $C(S)$ is a pro-C^*-algebra by Example 1.1.4(4), and the function $f(z) = z$ is an element of $C(S)$ for which $sp(f) = S$.

The spectrum does have a good description in terms of inverse systems. If $A = \lim\limits_{\longleftarrow d \in D} A_d$, where all the maps in the inverse system are unital, then an examination of coherent sequences shows that $sp((a_d)_{d \in D}) = \bigcup\limits_{d \in D} sp(a_d)$. If A is not unital, then one gets a similar result by looking at A^+. Examination of coherent sequences also proves the following list of analogs of standard facts about the spectrum and functional calculus in C^*-algebras.

1.2.2. PROPOSITION. *Let $A \cong \lim\limits_{\longleftarrow} A_d$ be a pro-C^*-algebra. If A is unital, we require that the image in each A_d of the unit of A be a unit for A_d. Let $x \in A$ correspond to the coherent sequence $(x_d)_{d \in D}$. Then:*

(1) For any open set U containing $sp(x)$, there is a holomorphic functional calculus $f \mapsto f(x)$ from the space of holomorphic functions on U (vanishing at 0 if A is not unital) to A. The coherent sequence corresponding to $f(x)$ is $(f(x_d))_{d \in D}$, and $sp(f(x)) = f(sp(x))$. The map $f \mapsto f(x)$ is continuous when the holomorphic functions are given the topology of uniform convergence on compact sets.

(2) If x is normal ($x^*x = xx^*$; this is the case if and only if each x_d is normal) then there is a continuous functional calculus $f \mapsto f(x)$ from $C(sp(x))$ (from $\{f \in C(sp(x)) : f(0) = 0\}$ if A is not unital) to A. The coherent sequence corresponding to $f(x)$ is $(f(x_d))_{d \in D}$, and $sp(f(x)) = f(sp(x))$. The map $f \mapsto f(x)$ is a homomorphism of pro-C^*-algebras.

(3) The following conditions are equivalent:

(3.1) x is selfadjoint ($x^* = x$).

(3.2) x is normal and $sp(x) \subset \mathbf{R}$.

(3.3) x_d is selfadjoint for all $d \in D$.

(4) The following conditions are equivalent:

(4.1) $x \geq 0$ (that is, x is selfadjoint and $sp(x) \subset [0, \infty]$).

(4.2) $x = y^2$ for some selfadjoint $y \in A$.

(4.3) $x = y^*y$ for some $y \in A$.

(4.4) $x_d \geq 0$ for all $d \in D$.

(5) *If A is unital, then the following conditions are equivalent:*

(5.1) *x is unitary (that is, $x^*x = xx^* = 1$).*

(5.2) *x is normal and $sp(x) \subset \{\lambda \in \mathbf{C} : |\lambda| = 1\}$.*

(5.3) *x_d is unitary for all $d \in D$.*

We now turn to the description of the algebra of bounded elements.

1.2.3. DEFINITION ([63]). Let A be a pro-C^*-algebra. Then an element $a \in A$ is called *bounded* if $\|a\|_\infty = \sup\{p(a) : p \in S(A)\} < \infty$. The set of all bounded elements is denoted $b(A)$.

1.2.4. PROPOSITION ([63]). *$b(A)$ is a C^*-algebra for the norm $\|\cdot\|_\infty$.*

Proof. The only nontrivial point is completeness. So let (a_n) be a Cauchy sequence in $b(A)$. Then (a_n) is Cauchy for each $p \in S(A)$, so there is $a \in A$ such that $p(a_n - a) \to 0$ for all $p \in S(A)$, by the completeness of A. Now let $\varepsilon > 0$. Choose N such that for $m, n \geq N$ we have $\|a_n - a_m\|_\infty < \frac{\varepsilon}{2}$. Then for $n \geq N$ and $p \in S(A)$, we have

$$p(a_n - a) = \lim_{m \to \infty} p(a_n - a_m) \leq \frac{\varepsilon}{2}.$$

Therefore $\|a_n - a\|_\infty \leq \frac{\varepsilon}{2} < \varepsilon$, and we have proved that $a_n \to a$ for $\|\cdot\|_\infty$. Q.E.D.

1.2.5. EXAMPLES. (1) If X is a compactly generated Hausdorff space, then $b(C(X)) = C_b(X)$, the algebra of all bounded continuous functions on X, with the supremum norm.

(2) If A_n is a C^*-algebra for each n, then $b(\prod_{n=1}^{\infty} A_n)$ is the C^*-algebra of all bounded sequences (a_n), again with the supremum norm.

These examples justify the term "bounded elements". They also show that, in general, the norm topology on $b(A)$ is much stronger than the restriction to $b(A)$ of the original topology on A. Thus, even in problems involving only bounded elements of a pro-C^*-algebra A, such as unitaries and projections, one cannot simply replace A by the C^*-algebra $b(A)$. For example, the unitary group of $C(\mathbf{R})$ is path-connected, but the unitary group of $C_b(\mathbf{R})$ is not connected, even though it is the same as a set. (See Example 1.2 in [54] for details.) Note also that every element of the unitary group of $C(\mathbf{R})$ has a logarithm in $C(\mathbf{R})$, but that many elements have no logarithms in $C_b(\mathbf{R})$. Also see the remarks following Proposition 1.5.3.

1.2.6. PROPOSITION. *Let A be a pro-C^*-algebra, and let $a \in A$ be normal. Then a is bounded if and only if $sp(a)$ is bounded.*

Proof. Let $\kappa_p : A \to A_p$ be the canonical map. Then

$$\|a\|_\infty = \sup_{p \in S(A)} \|\kappa_p(a)\| = \sup_{p \in S(A)} \Big(\sup_{\lambda \in sp(\kappa_p(a))} |\lambda| \Big) = \sup_{\lambda \in sp(a)} |\lambda|.$$

Q.E.D.

1.2.7. COROLLARY ([63]). *$b(A)$ is dense in A.*

Proof. It suffices to show that the selfadjoint elements of $b(A)$ are dense in the selfadjoint elements of A. Let $a \in A$ be selfadjoint, and let (f_n) be a sequence of bounded continuous functions such that $f_n(0) = 0$ and $f_n(\lambda) \to \lambda$ uniformly on compact subsets of $\mathbf{R}$. Then each $f_n(a)$ is bounded by the previous proposition, and $f_n(a) \to a$ by Proposition 1.2.2(3). Q.E.D.

1.2.8. COROLLARY ([63]). *If A is a pro-C^*-algebra and $p \in S(A)$, then the map $\kappa_p : A \to A_p$ is surjective. That is, $A/Ker(p)$ is complete.*

Proof. Since κ_p is continuous, the previous corollary implies that $\kappa_p(b(A))$ is dense in $\kappa_p(A)$ and hence in A_p. Since $b(A)$ and A_p are C^*-algebras, it follows that $\kappa_p|_{b(A)}$ is already surjective. Q.E.D.

The bounded elements also provide the link between pro-C^*-algebras as defined in [71] and pro-C^*-algebras as defined here. If A is a pro-C^*-algebra in the sense of this paper, then the pair $(b(A), S)$, where S is any cofinal subset of

$S(A)$, is a pro-C^*-algebra in the sense of [71]. Conversely, if (A_0, S) is a pro-C^*-algebra in the sense of [71], then $A = \varprojlim_{p \in S} A/\mathrm{Ker}(p)$ is a pro-C^*-algebra in our sense, for which S defines a cofinal subset of $S(A)$. One readily sees that these assignments define a category equivalence.

1.3. Generators and relations. Many of the most interesting pro-C^*-algebras in Part 2 of this article can be defined in terms of generators and relations. C^*-algebras defined in terms of generators and relations were introduced by Blackadar in [11], although some individual cases had been considered earlier, such as the irrational rotation algebras [60] (see page 417), the Cuntz algebras $\mathcal{O}_n$ [19] and $\mathcal{O}_A$ [23], and the noncommutative Grassmannians and unitary groups [16]. In order that a universal C^*-algebra on a given set of generators and relations exist, some fairly restrictive conditions must be satisfied. The advantage of allowing pro-C^*-algebras is that a universal pro-C^*-algebra on a set of generators and relations exists under much weaker hypotheses.

We begin with the construction of the C^*-algebra on a properly chosen set of generators and relations. Our treatment is based that of [11], but includes changes to make it more suitable for our purposes. If G is any set, we denote by $F(G)$ the free associative complex *-algebra (without identity) on the set G. Thus, $F(G)$ consists of all polynomials in the noncommuting variables $G \coprod G^*$ (disjoint union), with complex coefficients and no constant term. By definition, any function ρ from G to a C^*-algebra A extends to a unique *-homomorphism, which we also call ρ, from $F(G)$ to A.

A set R of relations on G is a collection of statements about the elements of G which make sense for elements of a C^*-algebra. Possible relations include statements of the form "$\|x\| \in S$", where $x \in F(G)$ and $S \subset \mathbf{R}$, "x is positive," or the statement that some equation in the variables $G \coprod G^*$ and some unknowns has a solution, or that some function from a topological space into G is continuous. Note that Blackadar considers only relations of the form $\|x\| \leq \eta$ for

$\eta \geq 0$ and x in the unitization $F(G)^+$ of $F(G)$. (We do not allow relations involving elements of $F(G)^+ - F(G)$ because they do not make sense in a nonunital C^*-algebra. However, it is perfectly possible for R to include the relations $eg = ge = g$ for some fixed $e \in G$ and all $g \in G \coprod G^*$.)

1.3.1. DEFINITION. Let (G, R) be a set of generators and relations. (That is, G is a set and R is a set of relations on G.) Then a *representation* of (G, R) in a C^*-algebra A is a function $\rho : G \to A$ such that the elements $\rho(g)$ for $g \in G$ satisfy the relations R in A. A representation on a Hilbert space H is a representation in $L(H)$.

1.3.2. DEFINITION. (Compare [11]). A set (G, R) of generators and relations is *admissible* if the following conditions hold:

(1) The function from G to the zero C^*-algebra is a representation of (G, R).

(2) If ρ is a representation of (G, R) in a C^*-algebra A, and if B is a C^*-subalgebra of A which contains $\rho(G)$, then ρ is a representation of (G, R) in B.

(3) If ρ is a representation of (G, R) in a C^*-algebra A, and if $\varphi : A \to B$ is a surjective homomorphism, then $\varphi \circ \rho$ is a representation of (G, R) in B.

(4) For every $g \in G$ there is a constant $M(g)$ such that $\|\rho(g)\| \leq M(g)$ for all representations ρ of (G, R).

(5) If $\{\rho_\alpha\}$ is a family of representations of (G, R) on Hilbert spaces H_α, then $g \mapsto \rho(g) = \bigoplus_\alpha \rho_\alpha(g)$ is a representation of (G, R) on $H = \bigoplus_\alpha H_\alpha$. (That is, the elements $\rho(g)$, which are in $L(H)$ by (4), in fact satisfy the relations R.)

Note that, in the presence of (3), condition (1) is equivalent to "there exists a representation of (G, R)." Also note that, for relations of the sort considered by Blackadar, (2) and (3) are automatic and (5) follows from (4).

1.3.3. EXAMPLES. (1) The relation $\|x\| = 1$ on the single generator x does not satisfy (1) of the definition.

(2) The relations "1 is an identity, $uu^* = u^*u = 1$, and there exists a continuous path $t \mapsto u_t$ with $u_0 = 1, u_1 = u$, and $u_t u_t^* = u_t^* u_t = 1$," on the generators 1 and u, satisfy (1) but not (2).

(3) The relation $\|x\| \in \{0, 1\}$ on the generator x satisfies (1) and (2) but not (3).

(4) The relation $x^* = x$ satisfies (1), (2), and (3), but not (4).

(5) The relations $x^* = x$ and $\|x\| < 1$ satisfy (1)-(4) but not (5).

(6) A more interesting (G, R) satisfying (1) through (4) but not (5) is $G = \{p_t : t \in [0, 1]\}$ and R consists of the relations $p_t^2 = p_t^* = p_t$ for all t, together with the relation $t \mapsto p_t$ is continuous.

(7) The relations $xp - px = 1$, $x \cdot 1 = 1 \cdot x = x$, $p \cdot 1 = 1 \cdot p = p$, and $1^* = 1^2 = 1$, on the generators $G = \{x, p, 1\}$, satisfy all of (1) through (5), but admit no nonzero representations.

(8) See Example 1.3 of [11] for a long list of admissible sets of generators and relations which admit nonzero representations.

The universal C^*-algebra on the generators G and relations R is a C^*-algebra $C^*(G, R)$ with a representation ρ of (G, R) in $C^*(G, R)$, such that, given any representation σ of (G, R) in a C^*-algebra B, then is a unique homomorphism $\varphi : C^*(G, R) \to B$ such that $\sigma = \varphi \circ \rho$. If (G, R) is admissible, then $C^*(G, R)$ exists and, following Blackadar, can be obtained as the Hausdorff completion of $F(G)$ in the C^*-seminorm

$$\|x\| = \sup\{\|\rho(x)\| : \rho \text{ is a representation of } (G, R).\}.$$

Note that condition (4) guarantees that $\|x\| < \infty$ for $x \in F(G)$, and that condition (5) guarantees that the obvious map from G to $C^*(G, R)$ is in fact a representation.

Admissibility, or something close to it, is also necessary for the existence of $C^*(G, R)$. Condition (1) is needed, since otherwise there may be no representations at all; conditions (2) and (3) are needed to ensure that the notion of a universal C^*-algebra is sensible, and without conditions (4) and (5) it will not be possible to construct a universal C^*-algebra. However, if the relations R also make sense in a pro-C^*-algebra, then a pro-C^*-algebra with the required properties will exist under much weaker conditions, as in the following definition. A representation of (G, R) in a pro-C^*-algebra has the obvious meaning.

1.3.4. DEFINITION. A set (G, R) of generators and relations is called *weakly admissible* if the following conditions are satisfied:

(1) The zero map from G to the zero C^*-algebra is a representation of (G, R).

(2) If ρ is a representation of (G, R) in a C^*-algebra A, and B is a C^*-subalgebra of A containing $\rho(G)$, then ρ is a representation of (G, R) in B.

(3) If ρ is a representation of (G, R) in a pro-C^*-algebra A, and $\varphi : A \to B$ is a surjective homomorphism to a C^*-algebra B, then $\varphi \circ \rho$ is a representation of (G, R) in B.

(4) If A is a pro-C^*-algebra, and $\rho : G \to A$ is a function such that, for every $p \in S(A)$, the composition of ρ with $A \to A_p$ is a representation of (G, R) in A_p, then ρ is a representation of (G, R).

(5) If $\rho_1, \ldots, \rho_n$ are representations of (G, R) in C^*-algebras $A_1, \ldots, A_n$, then $g \mapsto (\rho_1(g), \ldots, \rho_n(g))$ is a representation of (G, R) in $A_1 \oplus \cdots \oplus A_n$.

1.3.5. EXAMPLE. Any combination (including the empty set) of the following kinds of relations is weakly admissible:

(1) Any algebraic relation among the elements of $F(G)$, or the elements of the $n \times n$ matrix algebra $M_n(F(G))$.

(2) Any norm inequality of the form $\|x\| \leq \eta$ $(\eta \geq 0)$ or $\|x\| < \eta$ $(\eta > 0)$, where $x \in F(G)$ or $x \in M_n(F(G))$, and where norm relations are interpreted as applying to all continuous C^*-seminorms. (Thus, $\|x\| < 1$ means $p(x) < 1$ for all p. Note that this is a weaker condition than $\|x\|_\infty < 1$.)

(3) Any operator inequality $x \geq 0$ or $x \geq y$ for $x, y \in M_n(F(G))$.

(4) The assertion that a given function from an appropriate space to $F(G)$ is continuous, Lipschitz, differentiable, continuously differentiable, or r times (continuously) differentiable for $0 \leq r \leq \infty$.

1.3.6. PROPOSITION. *Let (G, R) be a weakly admissible set of generators and relations. Then there exists a pro-C^*-algebra $C^*(G, R)$, equipped with a representation $\rho : G \to C^*(G, R)$ of (G, R), such that, for any representation σ of (G, R) in a pro-C^*-algebra B, there is a unique homomorphism $\varphi : C^*(G, R) \to B$ satisfying $\sigma = \varphi \circ \rho$. If (G, R) is admissible, then $C^*(G, R)$ is a C^*-algebra.*

Proof. Let D be the set of all C^*-seminorms on $F(G)$ of the form $p(x) = \|\sigma(x)\|$ for some representation σ of G in a C^*-algebra. Then take $C^*(G, R)$ to be the Hausdorff completion of $F(G)$ in the family of C^*-seminorms D, with the obvious map $\rho : G \to C^*(G, R)$. To see that ρ is a representation, let $q \in S(C^*(G, R))$. Since D is a directed set in the obvious order (by (5) of the definition of weak admissibility), there is a representation σ of (G, R) in a C^*-algebra B such that $q(x) \leq \|\sigma(x)\|$ for all $x \in F(G)$. By (2) we may assume that B is generated by $\sigma(G)$, from which it follows that there is a surjective map $\varphi : B \to C^*(G, R)_q$ such that, with $\kappa_q : C^*(G, R) \to C^*(G, R)_q$ being the canonical map, we have $\varphi \circ \sigma = \kappa_q \circ \rho$. So $\kappa_q \circ \rho$ is a representation by (3), whence ρ is a representation by (4).

To show that $C^*(G, R)$ satisfies the universal property, it suffices, in view of (3), the definition of an inverse limit, and Corollary 1.2.8, to show the universal property holds for representations in C^*-algebras. Now use the definition of $C^*(G, R)$.

If (G, R) is admissible, then the directed set D has a largest element p, so that $C^*(G, R)$ is a C^*-algebra with the norm p. Q.E.D.

It follows that the sets of generators and relations in Examples 1.3.3 (4), (5), and (6) each define a universal pro-C^*-algebra. In Example 1.3.3 (4), it is

$\{f \in C(\mathbf{R}) : f(0) = 0\}$, and in Example 1.3.3 (5) it is $\{f \in C((-1,1)) : f(0) = 0\}$. In both cases, the generator x is the function $f(\lambda) = \lambda$.

1.3.7. REMARK. Let A be any pro-C^*-algebra. Then A can be realized as $C^*(G, R)$ for some weakly admissible set (G, R) of generators and relations. Indeed, we take $G = A$, and we take R to be the set of all algebraic relations which hold among the elements of A, together with the relation that the identity map from A to G is continuous. (Note that the continuity relation is necessary in general, because there exists a discontinuous bijective *-homomorphism of pro-C^*-algebras – see Example 1.4.9.)

We finish this section with two examples. The first shows that condition (4) in Definition 1.3.4 does not automatically follow from the other conditions. This is why we did not avoid the issue of whether relations make sense in pro-C^*-algebras by defining a representation of (G, R) in a pro-C^*-algebra A to be a function such that the maps $G \to A_p$ are representations for all $p \in S(A)$. The second example illustrates how, at least in some cases, the universal pro-C^*-algebra on a weakly admissible set of generators and relations can be obtained as an inverse limit of universal C^*-algebras on admissible sets of generator and relations.

1.3.8. EXAMPLE. Let A be the σ-C^*-algebra in Example 3.7 of [54], let $U(A)$ be the unitary group of A, and let $U_0(A)$ be the path-component of the identity in $U(A)$. It is shown in [54] that $U_0(A)$ is not closed in $U(A)$, so there exists $u \in U(A) - U_0(A)$ such that u is a limit of a net (u_α) in $U_0(A)$. Let $G = A$, and let R consist of all of the algebraic relations which hold among the elements of A, together with the relation that there exist a continuous path $t \mapsto u_t$ with $u_0 = 1, u_1 = u$, and $u_t u_t^* = u_t^* u_t = 1$ for all $t \in [0,1]$. Let $\rho : G \to A$ be the identity map. Then ρ is not a representation of (G, R). However, if $p \in S(A)$, and $\kappa_p : A \to A_p$ is the canonical map, then we have $\kappa_p(u_\alpha) \in U_0(A_p)$ and $\kappa_p(u_\alpha) \to \kappa_p(u)$, so that $\kappa_p(u) \in U_0(A_p)$

because $U_0(A_p)$ is closed. Thus, $\kappa_p \circ \rho$ is a representation of (G, R), and condition (4) of Definition 1.3.4 fails.

Conditions (1), (3), and (5) of the definition obviously hold. For condition (2), we note that, if R_0 consists just of the algebraic relations which hold in A, then $C^*(G, R_0) \cong A$ by Theorem 1.1.6. The argument at the end of the previous paragraph thus shows that the representations of (G, R) in C^*-algebras are exactly the same as the representations of (G, R_0) in C^*-algebras. Therefore (2) holds.

1.3.9. EXAMPLE. Let G and R be as in Example 1.3.3 (6), so that $C^*(G, R)$ is the universal pro-C^*-algebra on a continuous path $t \rightharpoonup p_t$, for $t \in [0, 1]$, of selfadjoint projections. Let D be the set of all nondecreasing functions $\delta : (0, \infty) \to (0, \infty]$, with the reverse pointwise order $\delta_1 \leq \delta_2$ if $\delta_1(\varepsilon) \geq \delta_2(\varepsilon)$ for all $\varepsilon > 0$. For $\delta \in D$ let R_δ consist of the relations "each p_t is a projection, and, for every $\varepsilon > 0$, we have $\|p_t - p_s\| \leq \varepsilon$ whenever $|t - s| \leq \delta(\varepsilon)$." Then each (G, R_δ) is admissible, and for $\delta_1 \geq \delta_2$ there is a canonical homomorphism of C^*-algebras $\pi_{\delta_1, \delta_2} : C^*(G, R_{\delta_1}) \to C^*(G, R_{\delta_2})$.

We claim that the obvious homomorphism

$$\varphi : C^*(G, R) \to \varprojlim_{\delta \in D} C^*(G, R_\delta)$$

is an isomorphism. It suffices to show that any representation ρ of (G, R) in a C^*-algebra is in fact a representation of (G, R_δ) for some $\delta \in D$. Set

$$\delta(\varepsilon) = \sup\{\delta_0 : \|\rho(p_t) - \rho(p_s)\| \leq \varepsilon \text{ whenever } |t - s| \leq \delta_0\}.$$

Then δ is nondecreasing, and maps $(0, \infty)$ to $(0, \infty]$ because $t \mapsto \rho(p_t)$ is uniformly continuous. So $\delta \in D$ and ρ is a representation of (G, R_δ).

1.4. **Commutative pro-C^*-algebras.** In this section, we give a description of the commutative unital pro-C^*-algebras which is analogous to the standard description of the commutative unital C^*-algebras. We then give some examples which demonstrate the inadequacy of earlier descriptions of the commutative unital pro-C^*-algebras, and which illustrate some of the pathologies that can occur in the category of pro-C^*-algebras.

It is unfortunately not true that the category of commutative unital C^*-algebras is equivalent to a nice category of topological spaces. (A mistake in an earlier proof to the contrary was pointed out to us by Jens Weidner.) Instead, one must consider spaces with distinguished families of compact subsets, as in the following definition, taken from Section 2 of [53]:

1.4.1. DEFINITION. Let X be a topological space. Then a family F of compact subsets is said to be *distinguished* if it contains all one point sets, is closed under finite unions and passage to compact subsets, and determines the topology in the sense that a subset C of X is closed if and only if $C \cap K$ is closed for all $K \in F$. If (X_1, F_1) and (X_2, F_2) are spaces with distinguished families of compact subsets, then a morphism from (X_1, F_1) to (X_2, F_2) is a continuous function $f : X_1 \to X_2$ such that $f(K) \in F_2$ for every $K \in F_1$.

If X is a topological space, and F is a set of subsets of X, then we write $C_F(X)$ for the topological *-algebra of all continuous functions from X to $\mathbf{C}$, with the topology of uniform convergence on the members of F. If F is omitted, it is understood to be the set of all compact subsets of F. Of course, in general $C_F(X)$ can fail to be a pro-C^*-algebra by not being complete.

We need one more definition.

1.4.2. DEFINITION. We call a topological space X *completely Hausdorff* if for any two distinct points $x, y \in X$ there is a continuous function $f : X \to [0,1]$ such that $f(x) = 0$ and $f(y) = 1$.

This condition lies between Hausdorff and completely regular. We will see in the examples that, even among the compactly generated spaces, it neither implies nor is implied by regularity.

1.4.3. THEOREM ([53]). *The assignment* $(X, F) \mapsto C_F(X)$ *is a contravariant category equivalence from the category of completely Hausdorff spaces with distinguished families of compact subsets to the category of commutative unital pro-C^*-algebras and unital homomorphisms.*

Notice that if (X, F) is as in this theorem then X is necessarily compactly generated; furthermore, every compactly generated space and every family of compact sets satisfying the closure conditions in Definition 1.4.1, and determining the topology, can occur. Previous descriptions of the commutative unital pro-C^*-algebras, such as Satz 1.1 of [63], have all been of the form $A \cong C_F(Y)$, where Y is completely regular and F is a family of compact subsets of Y satisfying the conditions of Definition 1.4.1 except that it is not required to determine the topology. These descriptions do not specify which completely regular spaces can occur, and, for a given completely regular space which can occur, they do not specify which families of compact subsets are allowed. In Example 2.1(3) of [36], it is suggested that any completely regular space X and any family F of compact sets, closed under finite unions and covering X, can occur. This, however, is false. In Example 1.4.6 below, we will produce a completely regular space X such that $C(X)$ is not complete in any topology whatever, and in Example 1.4.11, we will produce a compact space X and a family F of compact sets, of the sort appearing in previous descriptions of the commutative unital pro-C^*-algebras, such that $C_F(X)$ is not complete.

The previous descriptions of the commutative unital pro-C^*-algebras A were obtained as follows: let X be the set of (continuous) unital homomorphisms from A to $\mathbf{C}$, and give it the weak* topology. Since compactly generated completely Hausdorff spaces need not be regular (Example 1.4.7), and since Hausdorff topological vector spaces are completely regular ([40], 15.2 (3)), we see that the topology on a space Y resulting from its embedding in $C(Y)^*$ need not agree with the "right" topology on Y according to our Theorem 1.4.3. There is, however, a canonical procedure for recovering (X, F) from $C_F(X)$ which we now describe.

Let A be a commutative unital pro-C^*-algebra, and let X be the set of unital homomorphisms from A to $\mathbf{C}$. For any compact Hausdorff space K, let $Q(K, X)$ be the set of all functions $g : K \to X$ such that the formula $\varphi_g(a)(x) = g(x)(a)$ defines a (continuous, of course) homomorphism from A to $C(K)$. It can be shown that the sets $Q(K, X)$ define a quasitopology ([67]) on X, which is completely Hausdorff in the sense that for distinct $x, y \in X$ there is a "quasicontinuous" ([67]) function $f : X \to [0, 1]$ such that $f(x) = 0$ and $f(y) = 1$. The category of completely Hausdorff quasitopological spaces, with appropriate morphisms, is equivalent to the category of completely Hausdorff spaces with distinguished families of compact sets, so we have:

1.4.4. THEOREM. *The category of commutative unital pro-C^*-algebras and unital homomorphisms is contravariantly equivalent to the category of completely Hausdorff quasitopological spaces.*

For details see Section 2 of [53]. Quasitopological spaces were invented to resolve certain difficulties in homotopy theory, relating to the topology of function spaces, so their appearance here is consistent with our use of pro-C^*-algebras for certain aspects of "noncommutative homotopy theory" in Sections 2.5 and 2.6.

The difficulties discussed above disappear if one restricts to σ-C^*-algebras. Call a space X *countably compactly generated* if it is a direct limit of a countable direct system of compact spaces. (Such spaces are also called k_ω-spaces – see for example Section 2 of [42], or hemicompact spaces – see for example [5].)

1.4.5. THEOREM *The assignment $X \mapsto C(X)$ is a contravariant category equivalence from the category of countably compactly generated Hausdorff spaces and continuous functions to the category of commutative unital σ-C^*-algebras and unital homomorphisms. Furthermore, the original topology on such a space X is the same as the weak* topology it gets from its identification with the set of all unital homomorphisms from $C(X)$ to $\mathbf{C}$.*

Proof. The first statement is Proposition 5.7 of [53]. It is shown in the proof of that proposition that a countably compactly generated Hausdorff space is completely regular. The second statement now follows from the fact that, in a completely regular space, the weak topology determined by all continuous complex-valued functions is the same as the original topology. Q.E.D.

We now give the examples promised earlier. We generally refer to [53] for detailed proofs that they have the properties claimed for them.

1.4.6. EXAMPLE. Let $\beta\mathbf{N}$ be the Stone-Čech compactification of $\mathbf{N}$, let $x_0 \in \beta\mathbf{N} - \mathbf{N}$, and let $X = \mathbf{N} \cup \{x_0\}$. Then X is completely regular. However, X is not compactly generated; in fact, all compact subsets of X are finite. (An infinite compact subset of X would have to be a closed subset of $\beta\mathbf{N}$ containing infinitely many integers, and hence containing a copy of $\beta\mathbf{N}$. Therefore it would be uncountable.) It follows that $C(X)$ is not complete. In Example 2.12 of [53], it is shown that there is in fact *no* topology on $C(X)$ which makes it a pro-C^*-algebra.

In the following examples, ω denotes the first infinite ordinal, Ω denotes the first uncountable ordinal, and intervals $[\kappa, \lambda], [\kappa, \lambda)$, etc. are taken with respect to the ordinals.

1.4.7. EXAMPLE. Let $Y = [0, \Omega] \times [0, \omega] - \{(\Omega, \omega)\}$, and let X be the space obtained from Y by identifying the closed subset $\{\Omega\} \times [0, \omega)$ to a point. It is shown in Example 2.13 of [53] that X is compactly generated, completely Hausdorff, and not regular.

1.4.8. EXAMPLE. Let Y be as in the previous example, and let $Z = Y \times \mathbf{Z} \cup \{\pm\infty\}$, where $Y \times \mathbf{Z}$ gets the product topology, a neighborhood base at $+\infty$ consists of the sets $Y \times (\mathbf{Z} \cap [n, \infty)) \cup \{+\infty\}$, and a neighborhood base at $-\infty$ consists of the sets $Y \times (\mathbf{Z} \cap (-\infty, n]) \cup \{-\infty\}$, for $n \in \mathbf{Z}$. In Z, make the identifications

$$(\lambda, \Omega, 2k+1) \sim (\lambda, \Omega, 2k+2) \quad \text{and} \quad (\omega, \mu, 2k) \sim (\omega, \mu, 2k+1)$$

for $k \in \mathbf{Z}$, $\lambda \in [0, \omega)$, and $\mu \in [0, \Omega)$. Call the resulting space X, and give it the quotient topology. It is shown in Example 3 of Section VII.7 of [28] that X is regular but not completely Hausdorff (the points $+\infty$ and $-\infty$ cannot be separated by a continuous function), and it is shown in Example 2.14 of [53] that X is compactly generated. It follows from this that in the inverse limit $C(X) \cong \varprojlim_K C(K)$, where K runs through the compact subsets of X, every map in the inverse system is surjective, but not every map $C(X) \to C(K)$ is surjective.

1.4.9. EXAMPLE. Let $X = [0, \Omega)$. It is well known that every element of $C(X)$ is bounded and has a limit at Ω. Thus, there is a *-isomorphism $\varphi : C(X) \to C([0, \Omega])$, but φ is not continuous (although its inverse is). Furthermore, evaluation at Ω is a discontinuous *-homomorphism from $C(X)$ to $\mathbf{C}$. See Proposition 12.2 and the remark following it in [46].

1.4.10. EXAMPLE (Weidner). Let $X = [0, 1]$, and let F be the family of all countable closed subsets of $[0, 1]$ with only finitely many limit points. Then F is a distinguished family of compact subsets, relative to the usual topology. (Note that the sets of the form $\{x_n\} \cup \{x\}$, where $x_n \to x$, already determine the topology.) Therefore, by Theorem 1.4.3, $C_F(X)$ is a pro-C^*-algebra which is not isomorphic to $C(X)$. In Example 2.11 of [53], it is shown that $C_F(X)$ is not isomorphic to $C(Y)$ for any completely Hausdorff space Y. Thus, we cannot get away from distinguished families of compact subsets.

1.4.11. EXAMPLE. Let $X = [0, 1]$, and let F be the family of all finite subsets of X. Then F contains all one point subsets of X, and is closed under finite unions and passage to compact subsets. X is completely regular, and $C(X)$ is even a C^*-algebra, but $C_F(X)$ is not a pro-C^*-algebra, because it is not complete.

1.4.12. EXAMPLE. The space $\mathbf{Q}$ is countable (in particular, σ-compact) and compactly generated, but not countably compactly generated. Thus

$C(\mathbf{Q})$ is a pro-C^*-algebra but not a σ-C^*-algebra. See Example 5.8 of [53] for details.

1.4.13. EXAMPLE (Compare [72], page 83). Let X be as in Example 1.4.7, let $E_0 \subset X$ and $x \in X - E_0$ be a closed set and a point which cannot be separated by a continuous real-valued function, and let $E = E_0 \cup \{x\}$. Let $I = \{f \in C(X) : f|_E = 0\}$, which is a closed ideal in $C(X)$. Then $C(X)/I$, with its quotient topology, is homeomorphic to a subalgebra A of $C(E)$. (Note that E is compactly generated by [74], I.4.15.) This algebra is not all of $C(E)$, since the function $f : E \to \mathbf{C}$, given by $f(x) = 1$ and $f|_{E_0} = 0$, is not in A. However, A is dense in $C(E)$, because for any compact subset K of E and any continuous $g : K \to \mathbf{C}$, there is $f \in C(X)$ such that $f|_K = g$. It follows that $C(X)/I$ is not complete. (Of course, its completion is a pro-C^*-algebra.)

1.5. **Analogs of some constructions on C^*-algebras**. In this section, we discuss the analogs for pro-C^*-algebras of the following standard constructions in the theory of C^*-algebras: tensor products, free products, approximate identities, multiplier algebras, and quotients. The method in every case is to reduce to the C^*-algebra situation by using the inverse limit definition of a pro-C^*-algebra. The only item which gives any trouble is quotients, where we have to restrict to σ-C^*-algebras to obtain satisfactory results. We omit most of the proofs.

1.5.1. DEFINITION (Compare [53], Section 3). Let A and B be pro-C^*-algebras. We define

$$A \otimes_{\max} B = \varprojlim_{(p,q)\in S(A)\times S(B)} A_p \otimes_{\max} B_q$$

and

$$A \otimes_{\min} B = \varprojlim_{(p,q)\in S(A)\times S(B)} A_p \otimes_{\min} B_q.$$

We furthermore say that A is *nuclear* if A_p is nuclear for every $p \in S(A)$.

With this definition, $A \otimes_{\max} B$ and $A \otimes_{\min} B$ are appropriate completions of the algebraic tensor product of A and B. The two tensor products are the same whenever one of the factors is nuclear, and in this case we simply write $A \otimes B$. The algebras $C(X)$ are always nuclear, and, in fact, $C(X) \otimes B$ is naturally isomorphic to the algebra of continuous functions from X to B, with the topology of uniform convergence on each compact subset of X in each continuous C^*-seminorm on B. Also, whenever $\varphi : A \to C$ and $\psi : B \to C$ are homomorphisms of pro-C^*-algebras whose ranges commute, there exists a unique homomorphism $\varphi \otimes \psi : A \otimes_{\max} B \to C$ such that $(\varphi \otimes \psi)(a \otimes b) = \varphi(a)\psi(b)$ for $a \in A$ and $b \in B$. The proofs of these facts are straightforward, and most of them can be found in [53], Section 3.

1.5.2. DEFINITION (Compare [71], 1.9). Let A, B, and C be pro-C^*-algebras, and let $\varphi : C \to A$ and $\psi : C \to B$ be isomorphisms of C with closed subalgebras of A and B respectively. Then the *amalgamated free product* is

$$A *_C B = \varprojlim_{(p,q) \in S} A_p *_{C_{(p,q)}} B_q.$$

Here $S = \{(p,q) \in S(A) \times S(B) : p \circ \varphi = q \circ \psi\}$, $C_{(p,q)} = C_{p \circ \varphi}$ (which is the same as $C_{q \circ \psi}$), and the free product on the right hand side is taken in the category of C^*-algebras, as in [10] and [16].

1.5.3. PROPOSITION. *Let A, B, C, φ, and ψ be as in the definition.*

(1) *Given any pro-C^*-algebra Z and homomorphisms $\alpha : A \to Z$ and $\beta : B \to Z$ such that $\alpha \circ \varphi = \beta \circ \psi$, there exists a unique homomorphism $\lambda : A *_C B \to Z$ extending α and β in the obvious way.*

(2) *$A *_C B \simeq C^*(G, R)$, where the generators G are $A \coprod B$ (disjoint union) and the relations R are the algebraic relations among the elements of A and B, the relations $\varphi(c) = \psi(c)$ for $c \in C$, and the relations that the obvious maps from A and B to G are continuous.*

Proof. Part (2) follows immediately from part (1), so we need to prove only part (1). It is obvious for the case in which Z is a C^*-algebra, from the definition of $A *_C B$ and the universal property of the amalgamated free product of C^*-algebras. The general case of part (1) then follows from the definition of the inverse limit. Q.E.D.

In our language, the assertion in [71] that $A\bar{*}B \neq A * B$ in general becomes the assertion that $b(A *_C B) \neq b(A) *_C b(B)$ in general. This should not be surprising. We don't even have $b(A \otimes B) = b(A) \otimes b(B)$ in general, even if $A = C(X)$ for X compact or if A is a simple C^*-algebra. (See Section 3 of [53].)

An *approximate identity* in a pro-C^*-algebra is defined in the obvious way. A net (e_λ) in a pro-C^*-algebra A is an approximate identity if and only if its image in every A_p is an approximate identity. Approximate identities always exist – any approximate identity for $b(A)$ is one for A. For details, see Section 3 of [53].

1.5.4. DEFINITION. Let A be a pro-C^*-algebra. Then its multiplier algebra $M(A)$ is the pro-C^*-algebra $M(A) = \varprojlim_{p \in S(A)} M(A_p)$. Here, for $p \geq q$ the maps $M(A_p) \to M(A_q)$ of the multiplier algebras are defined as in Theorem 4.2 of [3], but they need not be surjective in general.

The multiplier algebra of a pro-C^*-algebra has properties analogous to those of the multiplier algebra of a C^*-algebra, including a strict topology, a description of $M(C(X) \otimes A)$ in terms of functions from X to $M(A)$ for compactly generated spaces X, and the following theorem.

1.5.5. THEOREM. (1) *Let B be a pro-C^*-algebra and let A be a closed *-subalgebra which contains an approximate identity for B. Then there is a natural inclusion $M(A) \to M(B)$.*

(2) *Let A and B be σ-C^*-algebras, let A have a countable approximate identity, and let $\varphi : A \to B$ be surjective. Then φ determines a surjective homomorphism $M(A) \to M(B)$.*

For the proof of the theorem and the remarks preceding it, see Section 3 and Theorem 5.11 of [53]. The theorem is proved by reduction to the corresponding results for C^*-algebras [3].

The multiplier algebra also has a direct description, not involving inverse limits: it is the algebra of all continuous multipliers of A as a topological algebra. It seems to be necessary to assume continuity here. For details, see Section 3 of [53].

We conclude this section by considering quotients.

1.5.6. PROPOSITION ([53], Corollary 5.4; [36]; [71]). *Let A be a σ-C^*-algebra, and let I be a closed *-ideal in A. Then A/I, with the quotient topology, is a σ-C^*-algebra.*

It is proved in [36] that closed ideals are automatically selfadjoint.

This proposition fails for general pro-C^*-algebras, since A/I need not be complete. (See Example 1.4.13.) In this case, one should presumably replace A/I by its completion. We will need only the following special case.

1.5.7 DEFINITION. Let A be a pro-C^*-algebra. Then the *abelianization* of A is the Hausdorff completion of A in the topology determined by all continuous C^*-seminorms which vanish on the closed ideal $[A, A]$ generated by all commutators $xy - yx$ for $x, y \in A$.

The abelianization can also be described as the completion of $A/[A, A]$ with respect to the quotient topology, or as the inverse limit $\varprojlim A_p/[A_p, A_p]$. It is the largest abelian quotient of A.

1.6. **The representable K-theory of σ-C^*-algebras**. In this section, we discuss K-theory for pro-C^*-algebras, extending the usual K-theory for C^*-algebras. (A good reference for the K-theory of C^*-algebras is [12].) The obvious approach, namely defining $K_0(A)$ to be the algebraic K_0-group for a unital pro-C^*-algebra A, turns out not to give the right group in applications. For example, $K_0(C(X)) \cong K^0(\beta X)$, where βX is the Stone-Čech compactification of X. (For more on why $K_0(A)$ is bad, see Section 4 of [54].) Here, we describe a theory, which we call representable K-theory and denote by RK_*, for which $RK_0(C(X))$ is isomorphic to the topologists' representable K-group $RK^0(X)$. The group $RK^0(X)$ is the group $[X, F]$ of homotopy classes of continuous maps from X to the space F of Fredholm operators on a separable infinite-dimensional Hilbert space. (The notation RK^* is taken from [39].) We have so far only been able to prove useful results about our theory for σ-C^*-algebras, so we restrict to them throughout this section. Proofs of the results stated in this section can be found in [54].

In his thesis [72], Jens Weidner has defined KK-theory for general pro-C^*-algebras A and B. His group $KK(\mathbf{C}, B)$ must agree with our group $RK_0(B)$, since there is a natural homomorphism $RK_0(B) \to KK(\mathbf{C}, B)$ which is an isomorphism if B is a C^*-algebra, and since both sides satisfy the Milnor $\varprojlim^1$-sequence. However, his definition of $KK(\mathbf{C}, B)$ is more complicated than our definition of $RK_0(B)$.

K-theory of inverse limits of Banach algebras has also already been considered by Mallios in [43] and [44], but only under the assumption that the algebras involved also be Q-algebras, that is, that their groups of invertible elements be open. As pointed out in [33] and proved, independently, in [53], a Q-pro-C^*-algebra is necessarily a C^*-algebra. Thus, the main results of [44] apply only to C^*-algebras. There are, however, algebras of the sort considered in [43], namely commutative inverse limits of Banach algebras, which are Q-algebras but not Banach algebras. An example is the algebra $\mathbf{C}[[z]]$ of all formal power series in one variable over $\mathbf{C}$, with the topology of pointwise convergence of the coefficients.

To motivate our definition of representable K-theory, we begin with the formula $RK^0(X) = [X, F]$. Since F is homotopy equivalent to the unitary group $U(Q)$ of the Calkin algebra Q, we obtain $[X, F] \cong [X, U(Q)]$, which is the same as the group of path components of $U(C(X) \otimes Q)$. Letting $U_0(A)$ denote the path-component of the identity in the unitary group $U(A)$ of a unital pro-C^*-algebra A, and abbreviating $U(A)/U_0(A)$ to $(U/U_0)(A)$, we obtain $RK^0(X) \cong (U/U_0)(C(X) \otimes Q)$. This suggests defining $RK_0(A) = (U/U_0)(A \otimes Q)$ for a unital pro-C^*-algebra A.

Unfortunately, Q is not nuclear, so tensor products with Q do not behave well. However, Mingo has shown ([49], Proposition 1.13) that, for any unital C^*-algebra A, we have $K_0(A) \cong (U/U_0)(Q(A))$, where $Q(A)$ is defined to be the outer stable multiplier algebra $M(K \otimes A)/(K \otimes A)$. Here K is the algebra of compact operator on a separable infinite dimensional Hilbert space. Note that $Q(\mathbf{C}) = Q$, and that $C(X) \otimes Q$ is a subalgebra of $Q(C(X))$ for X compact. The results of the previous section show that $Q(A)$ makes sense for σ-C^*-algebras (the difficulty for general pro-C^*-algebras is the quotient), so we can replace $A \otimes Q$ by $Q(A)$, and make the following definition.

1.6.1. DEFINITION ([54]). Let A be a unital σ-C^*-algebra. Then we define $RK_0(A)$ to be the abelian group $RK_0(A) = (U/U_0)(Q(A))$.

Of course, one must prove that $RK_0(A)$ really is abelian. We should also mention that the problem with the quotient $M(K \otimes A)/(K \otimes A)$ is not the most serious obstruction to using this definition for general pro-C^*-algebras. See the comments following Definition 1.1.2.

If $\varphi : A \to B$ is a unital homomorphism of σ-C^*-algebras, then, with the help of Theorem 1.5.5, we construct a homomorphism from $M(K \otimes A)$ to $M(K \otimes B)$, and hence a homomorphism $Q\varphi : Q(A) \to Q(B)$. From $Q\varphi$ we obtain a homomorphism $\varphi_* : RK_0(A) \to RK_0(B)$. If A is not unital, we can now define $RK_0(A)$ to be the kernel of the obvious homomorphism from $RK_0(A^+)$ to $RK_0(\mathbf{C})$, and we set $RK_i(A) = RK_0(A \otimes C_0(\mathbf{R}^i))$. Here $C_0(X)$ is the algebra of continuous complex-valued functions vanishing at infinity on the locally compact Hausdorff space X.

With these definitions, we obtain the following standard properties for representable K-theory.

1.6.2. THEOREM ([54]). *RK_0 and RK_1 are covariant homotopy invariant functors from σ-C^*-algebras to abelian groups satisfying the following properties:*

(1) (*Agreement with ordinary K-theory*). *If A is a C^*-algebra then $RK_i(A)$ is naturally isomorphic to $K_i(A)$.*

(2) (*Agreement with representable K-theory*). *If X is countably compactly generated and Hausdorff, then $RK_i(C(X))$ is naturally isomorphic to $RK^i(X)$.*

(3) (*Long exact sequence*). *If $0 \to I \to A \to B \to 0$ is a short exact sequence of σ-C^*-algebras, then there is a natural six term exact sequence*

$$\begin{array}{ccccc} RK_0(I) & \longrightarrow & RK_0(A) & \longrightarrow & RK_0(B) \\ \uparrow & & & & \downarrow \\ RK_1(B) & \longleftarrow & RK_1(A) & \longleftarrow & RK_1(I). \end{array}$$

(4) (*Bott periodicity*). *There is a natural isomorphism $RK_i(A \otimes C_0(\mathbf{R}^2)) \cong RK_i(A)$.*

(5) (*Stability*). *For any pro-C^*-algebra A and any Hilbert space H, $RK_i(K(H) \otimes A)$ is naturally isomorphic to $RK_i(A)$. (Here $K(H)$ is the C^*-algebra of all compact operators on H.)*

(6) (*Milnor $\varprojlim^1$-sequence*). *If $A \cong \varprojlim A_n$, and all maps $A_{n+1} \to A_n$ are surjective, then there is a natural short exact sequence*

$$0 \to \varprojlim{}^1 RK_{1-i}(A_n) \to RK_i(A) \to \varprojlim RK_i(A_n) \to 0.$$

We remark that the sequence (6) should be expected, in view of the similar sequences in [48]. Note that such sequences also appear for contravariant functors evaluated on direct limits – see [64], Section 7. Corollary 7.4 of [17] is a specific example of this, as one sees from the isomorphism $C(\varprojlim X_n) \cong \varinjlim C(X_n)$.

Our representable K-theory gives something new even in the commutative case. First we point out that if X is locally compact and σ-compact, hence countably compactly generated, then the groups $K^0(X)$ and $RK^0(X)$ are both defined, but are usually not isomorphic. Instead of obtaining them by applying two different functors, we obtain them by applying the same functor, namely RK_0, to two different algebras, namely $C_0(X)$ and $C(X)$. With this in mind, observe that if X is locally compact and Y is countably compactly generated, then $RK_0(C_0(X) \otimes C(Y))$ is defined, and can be regarded as the K-theory of $X \times Y$ with compact supports in the X direction and which is representable in the Y direction.

Finally, we mention that we call our theory representable K-theory because it generalizes representable K-theory for spaces, not because we claim it is a representable functor in the sense of algebraic topology. However, we believe that it is in fact a representable functor – see Section 2.5.

Part 2: Applications.

2.1. **Arveson's tangent algebra**. The construction of the tangent algebra of a C^*-algebra, in Section 5 of [7], appears to be the first use of pro-C^*-algebras in a problem originating in the theory of C^*-algebras. Here, we simplify and extend this construction.

To explain the problem, we need a definition.

2.1.1. DEFINITION. Let A and B be pro-C^*-algebras. Then a *derivation from A to B* is a pair (φ, δ), where $\varphi : A \to B$ is a homomorphism and $\delta : A \to B$ is a linear map satisfying $\delta(ab) = \delta(a)\varphi(b) + \varphi(a)\delta(b)$ and $\delta(a^*) = \delta(a)^*$ for all $a, b \in A$.

Of course, we are really interested in the case in which both A and B are C^*-algebras, but we will need the added generality below. Note that, if A and B are C^*-algebras, then δ is automatically bounded. (See page 252 of [7].)

Arveson wanted to find a derivation from a given C^*-algebra A to some other algebra which is universal in the sense of the following definition.

2.1.2. DEFINITION. Let A be a C^*-algebra. Then a unital (pro-)C^*-algebra TA, equipped with a derivation (ι, δ_0) from A to TA, is called a *tangent algebra* for A if, for any derivation (φ, δ) from A to any other C^*-algebra B, there exists a unique homomorphism $\psi : TA \to B$ such that $\varphi = \psi \circ \iota$ and $\delta = \psi \circ \delta_0$.

An explanation of the term tangent algebra is given in [7]. Also see the example at the end of this section. The tangent algebra is clearly unique if it exists. However, if the C^*-algebra A has any nontrivial derivations to other C^*-algebras, then no C^*-algebra will do for TA: one would have $\|\delta\| \leq \|\delta_0\|$ for any derivation (φ, δ), but $(\varphi, \lambda\delta)$ is also a derivation for any $\lambda \in \mathbf{R}$.

It is, however, possible to choose TA to be a σ-C^*-algebra. Arveson gives a complicated construction, involving tensor products and multilinear maps, but we can give a much simpler construction using the generator and relations techniques developed in Section 1.3. Let G be the union of two disjoint copies of A. We denote the elements of the first copy by $\iota(a)$ and the elements of the second copy by $\delta_0(a)$, for $a \in A$. Let R be the following set of relations on G: the elements $\iota(a)$ satisfy all the algebraic relations which hold among the elements of a; the map $a \mapsto \delta(a)$ is linear and preserves adjoints; and $\delta_0(ab) = \delta_0(a)\iota(b) + \iota(a)\delta_0(b)$ for all $a, b \in A$. Then (G, R) is clearly weakly admissible (compare Example 1.3.5), although it is not admissible, and there exists a universal pro-C^*-algebra $TA = C^*(G, R)$ by Proposition 1.3.6. It is obvious that $(TA, (\iota, \delta_0))$ is a tangent algebra; furthermore, it clearly satisfies the universal property of the definition even when B is allowed to be only a pro-C^*-algebra.

It remains to prove that TA is actually a countable inverse limit. To this end, let R_n be the set of relations R, together with the additional relations $\|\delta_0(a)\| \leq n\|a\|$ for each $a \in A$. Then (G, R_n) is in fact admissible, and hence $T_nA = C^*(G, R_n)$ is a C^*-algebra. There are obvious maps $TA \to T_nA$ and $T_{n+1}A \to T_nA$, and therefore there is an obvious map $TA \to \varprojlim_n T_n A$. We claim that this map is an isomorphism. To see this, it suffices to prove that every representation ρ of (G, R) is in fact a representation of (G, R_n) for some n. Obviously ρ determines a derivation $(\rho \circ \iota,\ \rho \circ \delta_0)$, and it suffices to choose $n \geq \|\rho \circ \delta_0\|$. So TA is a σ-C^*-algebra.

The uniqueness of TA implies that our construction gives an algebra isomorphic to the one constructed by Arveson. Also note that our construction applies to pro-C^*-algebras: one need only add to R the relation that ι (and perhaps δ_0) be continuous.

The algebra TA contains information about all derivations whose domain is A. The most interesting derivations, however, are unbounded derivations defined on appropriate dense subalgebras of A. In [7], Arveson suggests extending the functor $A \mapsto TA$ to a category which would include some of these dense subalgebras, when they are endowed with their own natural topologies. For instance, this category should include $C^\infty(M)$, with the C^∞ topology, for a compact manifold M. Here, we propose a different approach. Let A be a C^*-algebra, and let A_0 be a dense *-subalgebra of A. Let G consist of the elements $\iota(a)$ for $a \in A$ and elements $\delta_0(a)$ for $a \in A_0$, and let R consist of the relations that ι be a *-homomorphism, that δ_0 be linear and preserve adjoints on A_0, and that $\delta_0(ab) = \delta_0(a)\iota(b) + \iota(a)\delta_0(b)$ for $a, b \in A_0$. Then (G, R) is again weakly admissible, and therefore admits a universal pro-C^*-algebra $C^*(G, R)$, which we denote by $T(A, A_0)$. By construction, $(T(A, A_0), (\iota, \delta_0))$ satisfies the universal property in Definition 2.1.2, but where the derivations are assumed to be defined only on A_0. Of course, this construction also admits variants – for example, δ_0 can be required to be continuous for some appropriate topology on A_0. Also, getting a σ-C^*-algebra will presumably depend on A_0 having some good properties.

2.1.3. EXAMPLE. Let M be a compact smooth manifold (without boundary), let $A = C(M)$, and let $A_0 = C^\infty(M)$. We will show that the abelianization C of $T(A, A_0)$ (Definition 1.5.7) is canonically isomorphic to the σ-C^*-algebra $C(TM)$, where TM is the tangent space of M. This provides some real justification for calling $T(A, A_0)$ a tangent algebra. ($T(A, A_0)$ itself cannot be commutative, because it must be universal for all inner derivations of A into noncommutative C^*-algebras containing A as a subalgebra. See the remark about $T(C(X))$ on page 266 of [7].)

The abelianization C is clearly isomorphic to $C^*(G, R_0)$, where G is the set of generators used in the construction of $T(A, A_0)$, and R_0 is the set of relations used there together with the relations that the elements of $G \coprod G^*$ all commute. There is an obvious derivation (ι, δ_0) from $(C(M), C^\infty(M))$ to $C(TM)$, given by $\iota(f) = f \circ p$, where $p : TM \to M$ is the bundle projection map, and $\delta_0(f) = Df$, the derivative of f regarded as a map from TM to $\mathbf{C}$. To show that $C \cong C(TM)$, it suffices to show that the ranges of ι and δ_0 together generate $C(TM)$ as a pro-C^*-algebra, and that every derivation from $(C(M), C^\infty(M))$ to a commutative C^*-algebra factors through $C(TM)$.

To prove the first part of this, it is enough to show that for every compact set $K \subset TM$, the restriction to K of the ranges of ι and δ_0 generates $C(K)$. This is true because the range of ι separates any two points in distinct fibers of TM and the range of δ_0 separates any two points in the same fiber of TM.

For the second part, let (φ, δ) be a derivation from $(C(M), C^\infty(M))$ to $C(X)$ for some compact space X. (We can deal with the nonunital commutative C^*-algebras by unitizing them.) For $x \in X$, let ev_x be evaluation at x. Then $(ev_x \circ \varphi, ev_x \circ \delta)$ is a point derivation of $C^\infty(M)$, and therefore has the form $(ev_x \circ \varphi)(f) = f(m)$ and $(ev_x \circ \delta)(f) = Df(v)$ for some $m \in M$ and $v \in T_m M$, by Theorem 3 and the remarks following it on pages 107-109 of [68]. We thus obtain a function $h : X \to TM$ such that $\varphi(f) = f \circ p \circ h$ and $\delta_0(f) = Df \circ h$. Reasoning similar to that of the previous paragraph shows that h is continuous, and the proof that the abelianization of $T(C(M), C^\infty(M))$ is $C(TM)$ is complete.

This example deserves two remarks. First, Fred Goodman has shown us how to prove that if (φ, δ) is any derivation from $(C(M), C^\infty(M))$ to a C^*-algebra B, then δ is continuous in the C^k topology on $C^\infty(M)$ for some k. It follows that $T(C(M), C^\infty(M))$ is a σ-C^*-algebra, so that the map $T(C(M), C^\infty(M)) \to C(TM)$ is actually surjective. Secondly, the algebra $T(C(M), C^k(M))$ is much more complicated, since there exist many discontinuous derivations on $C^k(M)$. See Example 6.5 of [24] and the references cited there.

2.2. The multipliers of Pedersen's ideal. In [50], Pedersen constructed a minimal dense hereditary ideal K_A of a (nonunital) C^*-algebra A. This ideal was later shown to be minimal among all dense ideals in A. (Compare [51], 5.6.1.) In [41], Lazar and Taylor made an extensive study of the multiplier algebra $\Gamma(K_A)$ of K_A. Although they did not realize it at the time, $\Gamma(K_A)$ has a natural topology in which it is a pro-C^*-algebra, and a significant number of their results follow from general properties of pro-C^*-algebras. One can in fact give a very concrete description of $\Gamma(K_A)$ as an inverse limit of C^*-algebras. The results described in this section are proved in [56].

A multiplier of K_A is a pair (S, T) of linear maps from K_A to itself such that $S(xy) = S(x)y, T(xy) = xT(y)$, and $xS(y) = T(x)y$ for $x, y \in K_A$. No continuity condition is imposed. (The continuous multipliers of K_A are exactly the elements of multiplier algebra $M(A)$, which is a C^*-algebra.) Early in [41], Lazar and Taylor prove that if $a \in K_A$, then the closed left and right ideals $\overline{Aa}$ and $\overline{aA}$ generated by a are contained in K_A. Furthermore, they show that if $(S, T) \in \Gamma(K_a)$, then $S(\overline{Aa}) \subset \overline{Aa}$ and $T(\overline{a^*A}) \subset \overline{a^*A}$, and that the set

$$\begin{aligned} M_a = \{(S, T) : \quad & S : \overline{Aa} \to \overline{Aa} \text{ and } T : \overline{a^*A} \to \overline{a^*A} \\ & \text{are linear and satisfy } yS(x) = T(y)x \text{ for } x \in \overline{Aa} \text{ and } y \in \overline{a^*A}\} \end{aligned}$$

is a C^*-algebra in the operator norm $\|(S,T)\| = \|S\|$. (In particular, S must be bounded.) It is then essentially immediate that $\Gamma(K_A)$ is isomorphic to the pro-C^*-algebra $\varprojlim_{a \in (K_A)_+} M_a$, where $(K_A)_+$ is the set of positive elements of K_A, ordered in the usual way.

The results of Part 1 can now all be applied to $\Gamma(K_A)$. In particular, one immediately obtains holomorphic functional calculus for arbitrary elements and continuous functional calculus for normal elements of $\Gamma(K_A)$. In fact, the construction of functional calculus in pro-C^*-algebras (Proposition 1.2.2) is essentially trivial, while Lazar and Taylor went to some trouble to get results which, for the most part, are slightly weaker. We remark that the continuous functional calculus in $\Gamma(K_A)$ has been used by Rieffel in connection with the study of the ideal structure of crossed products by finite groups – it plays a key role in the proof of the main lemma of [59].

It is actually possible to identify the algebras M_a more concretely, at least if $a \geq 0$. Namely, we prove in [56] that $M_a \cong M(\overline{AaA})$, the multiplier algebra of the closed ideal generated by a. From this it is easy to prove that if the primitive ideal space Prim(A) of A is compact, then $\Gamma(K_A) = M(A)$, that is, all multipliers of K_A are bounded. (This result is also in [41], but with a much more complicated proof.) One then obtains the following concrete description of $\Gamma(K_A)$.

2.2.1. THEOREM ([56]). *Let A be a C^*-algebra, and let S be the set of all closed ideals I in A such that Prim(A/I) is compact, ordered by reverse inclusion. Then there are canonical isomorphisms*

$$\Gamma(K_A) \cong \varprojlim_{I \in S} M(A/I) \cong M(\varprojlim_{I \in S} A/I).$$

If A has a countable approximate identity, then each map $\Gamma(K_A) \to M(A/I)$ is surjective.

This theorem identifies $\Gamma(K_A)$ with the algebra of continuous multipliers of a certain pro-C^*-algebra, as defined in Section 1.5. The strict topology on this multiplier algebra turns out to agree approximately with the κ-topology on $\Gamma(K_A)$ used in [41]. Also, it should be pointed out that this theorem enables one to derive the Dauns-Hofmann theorem of [41] directly from the usual Dauns-Hofmann theorem. See [56] for details.

2.3. The K-theory of crossed products by homotopic actions. In this section, we consider the following question: if α and β are homotopic actions of a locally compact group G on a C^*-algebra A, what can one say about the K-theory of the crossed products by α and β? For $G = \mathbf{Z}$, $\mathbf{R}$, or F_n, known results easily imply that the K-theory of the two crossed products is the same. For $G = \mathbf{Z}/2\mathbf{Z}$, or more generally for G compact Lie, this is not true; nevertheless, something can be said, and the proof uses the representable K-theory of σ-C^*-algebras sketched in Section 1.6.

Homotopy of actions is defined as follows.

2.3.1. DEFINITION. A homotopy $t \mapsto \alpha^{(t)}$ of actions of a group G on a C^*-algebra A is a function from $[0, 1]$ (or other appropriate interval) to the set of actions of G on A such that, for each $a \in A$, the function $(t, g) \mapsto \alpha_g^{(t)}(a)$ from $[0, 1] \times G$ to A is jointly continuous.

It is easily seen that $t \mapsto \alpha^{(t)}$ is a homotopy if and only if the formula $\alpha_g(f)(t) = \alpha_g^{(t)}(f(t))$ defines an action α of G on the C^*-algebra $C([0, 1], A)$ of all continuous functions $f : [0, 1] \to A$. Throughout this section, whenever $t \mapsto \alpha^{(t)}$ is a homotopy, α will be understood to denote this action.

If $G = \mathbf{Z}$, and $t \mapsto \alpha^{(t)}$ is a homotopy of actions on A, then it is easily shown that $K_*(C^*(G, A, \alpha^{(t)}))$ is independent of t. The key point is that if γ is an action of $\mathbf{Z}$ on a contractible C^*-algebra C, then the Pimsner-Voiculescu exact sequence [57] shows that $K_*(C^*(\mathbf{Z}, C, \gamma)) = 0$. Note that the contraction of C is *not* assumed to be equivariant; all we really need to know is that $K_*(C) = 0$. Since formation of full crossed products preserve exactness ([52], Lemma 2.8.2), we have an exact sequence

$$0 \to C^*(\mathbf{Z},\ \mathrm{Ker}(ev_t), \alpha) \to C^*(\mathbf{Z}, C([0,1], A), \alpha) \to C^*(\mathbf{Z}, A, \alpha^{(t)}) \to 0,$$

where ev_t is evaluation at t. Since $\mathrm{Ker}(ev_t)$ is contractible, we obtain

$$K_*(C^*(\mathbf{Z}, A, \alpha^{(t)})) \cong K_*(C^*(\mathbf{Z}, C([0,1], A), \alpha))$$

for all t, and the right hand side does not depend on t. (In [29], Elliott solves essentially the same problem in a slightly different way.) Substituting for the Pimsner-Voiculescu exact sequence the Connes isomorphism [18] if $G = \mathbf{R}$, or the Pimsner-Voiculescu exact sequence for F_n [58] and the K-amenability of F_n [20] if $G = F_n$, we obtain the same result. For details see Proposition 3.2 of [55].

Unfortunately it is not in general true that the K-theory of the crossed product of a contractible C^*-algebra by an arbitrary action is zero, and in fact the K-theory of crossed products by homotopic actions can be different.

2.3.2 EXAMPLE ([55]). Let $G = \mathbf{Z}/2\mathbf{Z}$, let A be the 2^∞ UHF algebra $A = \bigotimes_1^\infty M_2(\mathbf{C})$, and let α be the product type action sending the nontrivial element g of G to $\alpha_g = \bigotimes_1^\infty \mathrm{ad}\left(\begin{smallmatrix} 1 & 0 \\ 0 & -1 \end{smallmatrix}\right)$. Then $C^*(G, A, \alpha) \cong A$, so that $K_*(C^*(G, A, \alpha)) \cong \mathbf{Z}[\frac{1}{2}]$, that is, the subgroup $\{\frac{k}{2^n} : k \in \mathbf{Z},\ n \in \mathbf{N} \cup \{0\}\}$ of $\mathbf{Q}$. We will sketch a proof that α is homotopic to the trivial action ι, for which we have $C^*(G, A, \iota) \cong A \oplus A$ and $K_*(C^*(G, A, \iota)) \cong \mathbf{Z}[\frac{1}{2}] \oplus \mathbf{Z}[\frac{1}{2}]$.

The first stage in the construction of the required homotopy is to find a path of order two automorphisms connecting α_g to $\mathrm{id}_{M_2(\mathbf{C})} \otimes \bigotimes_2^\infty \mathrm{ad}\left(\begin{smallmatrix} 1 & 0 \\ 0 & -1 \end{smallmatrix}\right)$. It clearly suffices to find a path of order two unitaries in $M_2(\mathbf{C}) \otimes M_2(\mathbf{C})$ connecting $u_0 = \left(\begin{smallmatrix} 1 & 0 \\ 0 & -1 \end{smallmatrix}\right) \otimes \left(\begin{smallmatrix} 1 & 0 \\ 0 & -1 \end{smallmatrix}\right)$ to $u_1 = \left(\begin{smallmatrix} 1 & 0 \\ 0 & 1 \end{smallmatrix}\right) \otimes \left(\begin{smallmatrix} 1 & 0 \\ 0 & -1 \end{smallmatrix}\right)$. Now $M_2(\mathbf{C}) \otimes M_2(\mathbf{C}) \cong M_4(\mathbf{C})$, and, under this isomorphism, u_0 and u_1 both become diagonal matrices with two $+1$ eigenvalues and two -1 eigenvalues. Therefore there is a permutation matrix w such that $w u_0 w^* = u_1$. If w_t is a unitary path connecting 1 to w, then the required path is $t \mapsto w_t u_0 w_t^*$. The second stage of the construction uses the same method to connect $\mathrm{id}_{M_2(\mathbf{C})} \otimes \bigotimes_{2}^{\infty} \mathrm{ad}\begin{pmatrix} 1 & 0 \\ 0 & -1 \end{pmatrix}$ to $\mathrm{id}_{M_2(\mathbf{C})} \otimes \mathrm{id}_{M_2(\mathbf{C})} \otimes \bigotimes_{3}^{\infty} \mathrm{ad}\begin{pmatrix} 1 & 0 \\ 0 & -1 \end{pmatrix}$, leaving the first factor alone. This process is repeated infinitely often, and yields a path of order two automorphisms from α_g to the limit at infinity, which can be shown to be id_A.

More details and some generalizations of this example appear in Section 3 of [55]. We also give there a similar example involving an action of $\mathbf{Z}/2\mathbf{Z}$ on a commutative unital C^*-algebra, originally suggested to us by Graeme Segal and based on the example in Section 5 of [9].

Nevertheless, something can be said about the K-theory of crossed products by homotopic actions of compact Lie groups, and it is at this point that it becomes necessary to introduce pro-C^*-algebras. The relevant theorems are both direct generalizations to the noncommutative case of results of [9]. Before stating them we recall some facts about the universal free G-space EG and about equivariant K-theory.

For the remainder of this section, G is a compact Lie group. The space EG is a free contractible G-space. It is of course not unique, but it is unique up to G-homotopy equivalence. It cannot be chosen to be compact, but it can be chosen to be countably compactly generated, for instance using the model of [47], topologized as in Section 2 of [9]. For example, if $G = S^1$, the circle group, then this gives for EG the unit sphere in $\varinjlim_n \mathbf{C}^n$, with the action given by scalar multiplication. Equivariant K-theory is defined in Chapter 2 of [52]; it is the noncommutative version of the theory in [66]. The key result, originally due to Julg [37], is the isomorphism $K_*^G(A) \cong K_*(C^*(G, A))$. This isomorphism makes $K_*(C^*(G, A))$ into a module over the representation ring $R(G) = K_0(C^*(G))$ of G, studied in [65]. We let $I(G)$ be the augmentation ideal in $R(G)$ (see the example after Proposition 3.7 of [65]), and, for any $R(G)$-module M, we denote its $I(G)$-adic completion (see Chapter 10 of [8]) by $M^\wedge$.

We can now state the theorems. In them, the compact Lie group G acts on the C^*-algebra A.

2.3.3. THEOREM (Atiyah-Segal completion theorem for C^*-algebras [55]). *Under certain technical hypotheses, there is a natural isomorphism*

$$K_*(C^*(G,A))^\wedge \cong RK_*([A \otimes C(EG)]^G).$$

2.3.4. THEOREM ([55]). *The group $RK_*([A \otimes C(EG)]^G)$ depends only on the homotopy type of the action of G on A.*

Here RK_* is the representable K-theory discussed in Section 1.6. The superscript G denotes the algebra of G-fixed points under the diagonal action, so that $[A \otimes C(EG)]^G$ is the noncommutative analog of the "homotopy quotient" $(X \times EG)/G$. The technical hypotheses in the first theorem are satisfied whenever $K_*(C^*(G,A))$ is finitely generated as an $R(G)$-module, and also whenever A is an AF algebra, G is abelian, and G leaves invariant an increasing sequence of finite dimensional subalgebras with dense union. For details see [55]. The proof of Theorem 2.3.3 is essentially the same as the proof of the original Atiyah-Segal completion theorem in [9], once the basic properties of representable K-theory for σ-C^*-algebras are known. The proof of Theorem 2.3.4 involves the generalization to bundles of C^*-algebras of results of Dold [27] on fiber homotopy equivalences.

2.3.5. COROLLARY ([55]). *Let α and β be two homotopic actions of G on A, both satisfying the technical hypotheses of Theorem 2.3.3. Then $K_*(C^*(G,A,\alpha))^\wedge \cong K_*(C^*(G,A,\beta))^\wedge$.*

It is shown in [9] that the technical hypotheses in Theorem 2.3.3 cannot be dropped, even when A is commutative and unital. However, we have no example to show that they cannot be dropped in this corollary.

2.3.6. EXAMPLE. Let A_θ be the irrational rotation algebra, $A_\theta = C^*(\mathbf{Z}, C(S^1), \beta)$, where $\theta \in [0,1] - \mathbf{Q}$ is fixed and $\beta(f)(z) = f(e^{-2\pi i\theta}z)$ for $f \in$

$C(S^1)$ and $z \in S^1$. Let α be the action of $\mathbf{Z}/n\mathbf{Z}$ on A_θ given by restricting the dual action $\hat{\beta}$ to the obvious copy of $\mathbf{Z}/n\mathbf{Z}$ in $\hat{\mathbf{Z}}$. Since $\hat{\mathbf{Z}} = S^1$ is path-connected, each α_g is homotopic to id_{A_θ}. However, in Example 4.5 of [55], it is shown to follow easily from Corollary 2.3.5 that α is *not* homotopic to the trivial action.

2.4. Noncommutative analogs of classical Lie groups. The irrational rotation algebras have long been known as "noncommutative tori," and more recently higher-dimensional "noncommutative tori" have been studied ([29], [25], [22], [61]). Brown introduced "noncommutative Grassmannians" and "noncommutative unitary groups" in [16]. Inspired no doubt by the success of these notions, and motivated by considerations arising from his results on the spectrums of elements of the reduced C^*-algebras of free groups, Voiculescu constructed in [71] noncommutative analogs of a number of other classical Lie groups. These algebras are of course really noncommutative analogs of the (commutative) algebras of continuous functions on tori, Grassmannians, unitary groups, etc. What is new in the cases considered by Voiculescu is that the original spaces are not compact. For reasons that will become clear below, the noncommutative analog of a Lie group G should be a noncommutative version of $C(G)$ and not of $C_0(G)$. Thus, the objects we will be looking at are not C^*-algebras; they are, however, σ-C^*-algebras.

In this section we will motivate and discuss the "noncommutative classical Lie groups" constructed in [71], but using our version of pro-C^*-algebras rather than Voiculescu's. (The two versions are of course equivalent, as discussed at the end of Section 2.) We begin by considering the "noncommutative unitary group" $U_{nc}(n)$ defined in [16]. (The subscript nc stands for "noncommutative".) It is the unital C^*-algebra with generators 1 and x_{ij} $(1 \leq i, j \leq n)$, subject to the relations that 1 be the identity and that the $n \times n$ matrix $(x_{ij})_{i,j=1}^n$ be a unitary element of $M_n(U_{nc}(n))$. This set of generators and relations is clearly admissible (Definition 1.3.2), since every x_{ij} must satisfy $\|x_{ij}\| \leq 1$. Therefore $U_{nc}(n)$ is in fact a C^*-algebra. Furthermore, the group structure on the usual unitary group $U(n)$

is also present in this algebra, in the sense that $U_{nc}(n)$ is a "dual group" in the category of unital C^*-algebras. This notion is defined as follows.

2.4.1. DEFINITION (Compare [71], 2.1.). A *dual group in the category of unital C^*-algebras* is a quadruple (A, μ, ι, χ) consisting of a unital C^*-algebra A and unital homomorphisms $\mu : A \to A *_{\mathbf{C}} A$, $\iota : A \to A$, and $\chi : A \to \mathbf{C}$, satisfying certain conditions. To state them, we introduce the unital homomorphisms $\varepsilon_A : \mathbf{C} \to A$, given by $\varepsilon(\lambda) = \lambda \cdot 1$, and $\delta_A : A *_{\mathbf{C}} A \to A$, mapping each copy of A in the free product identically onto A. Then the conditions are:

(1) $(\mu * \mathrm{id}_A) \circ \mu = (\mathrm{id}_A * \mu) \circ \mu$ (as maps from A to $A *_{\mathbf{C}} A *_{\mathbf{C}} A$).

(2) $\iota^2 = \mathrm{id}_A$.

(3) $\delta_A \circ (\iota * \mathrm{id}_A) \circ \mu = \delta_A \circ (\mathrm{id}_A * \iota) \circ \mu = \varepsilon_A \circ \chi$.

(4) $(\chi * \mathrm{id}_A) \circ \mu = (\mathrm{id}_A * \chi) \circ \mu = \mathrm{id}_A$, where $\mathbf{C} *_{\mathbf{C}} A$ and $A *_{\mathbf{C}} \mathbf{C}$ are identified in the obvious way with A.

To understand this definition, imagine that it is taking place in the category of sets, with the directions of all maps reversed, and with all category concepts replaced by their duals. This last operation means here that the free product $A *_{\mathbf{C}} A$ must be replaced by the product, and that the algebra $\mathbf{C}$ (which has a unique unital homomorphism ε_A to any unital C^*-algebra A) must be replaced by a one point set $\{e\}$ (which has a unique map from every other set). Rearranged in this manner, this definition simply says that A is a group, with multiplication $\mu : A \times A \to A$, inversion $\iota : A \to A$, and identity $\chi(e) \in A$. For this reason, the object we have defined is also called a "group object in the opposite category to the category of unital C^*-algebras".

We now return to $U_{nc}(n)$. We take $U_{nc}(n) *_{\mathbf{C}} U_{nc}(n)$ to be generated by 1 and $y_{ij}, z_{ij} (1 \leq i, j \leq n)$, where (y_{ij}) and (z_{ij}) are $n \times n$ unitary matrices. We then define μ by setting $\mu(x_{ij})$ equal to the ij entry of the matrix product $(y_{ij}) \cdot (z_{ij})$. We further set $\iota(x_{ij}) = x_{ji}^*$ and $\chi(x_{ij}) = \delta_{ij}$. Then it is readily verified that $(U_{nc}(n), \mu, \iota, \chi)$ is in fact a dual group. Its abelianization (see Definition 1.5.7) is $C(U(n))$, and the maps $U_{nc}(n) \to C(U(n)) \to \mathbf{C}$, where the last map is evaluation at $u \in U(n)$, are the unique maps sending x_{ij} to $u_{ij} \in \mathbf{C}$. Furthermore, the abelianizations of the maps μ, ι, and χ are the maps on the algebras of continuous functions corresponding to the group operations $U(n) \times U(n) \to U(n)$, $U(u) \to U(n)$, and $\{e\} \to U(n)$. This is the real justification for calling $U_{nc}(n)$ a "noncommutative unitary group."

Voiculescu wanted to do the same thing for certain noncompact Lie groups G. First, note that if G is locally compact but not compact, then the group multiplication $G \times G \to G$ is not a proper map, and therefore does not define a homomorphism from $C_0(G)$ to $C_0(G \times G)$. It does, however, define a homomorphism from $C(G)$ to $C(G \times G)$. Thus, if A is to be a noncommutative analog of G, then we must expect the abelianization of A to be $C(G)$, so that A will be a pro-C^*-algebra but not a C^*-algebra. A dual group in the category of unital pro-C^*-algebras is defined simply by replacing the word "C^*-algebra" by the word "pro-C^*-algebra" in Definition 2.4.1. This is in fact how this definition appears in [71]. The amalgamated free products appearing in the definition are defined in Section 1.5.

As an example, we will now define $GL_{nc}(n, \mathbf{C})$ ([71], 5.4). It is the universal pro-C^*-algebra on generators 1, x_{ij}, y_{ij} $(1 \leq i, j \leq n)$ satisfying the relations that 1 be an identity and that the $n \times n$ matrix (y_{ij}) be the inverse of the $n \times n$ matrix (x_{ij}). Note that these relations are weakly admissible but not admissible. The homomorphisms μ and χ are defined on the generators x_{ij} just as for $U_{nc}(n)$; the definitions on the y_{ij} are then forced. The homomophism ι is given by $\iota(x_{ij}) = y_{ij}$ and $\iota(y_{ij}) = x_{ij}$. Voiculescu shows that $GL_{nc}(n, \mathbf{C})$ is in fact a σ-C^*-algebra, by obtaining it as the inverse limit over k of the C^*-algebras in which the additional relations

$$k^{-1} \leq (x_{ij})^*(x_{ij}) \leq k \quad \text{and} \quad k^{-1} \leq (x_{ij})(x_{ij})^* \leq k$$

in the algebra of $n \times n$ matrices are required to hold.

In a similar but slightly more complicated manner, Voiculescu also defines dual groups $GL_{nc}(n, \mathbf{R})$, $U_{nc}^*(2n)$, $U_{nc}(p,q), O_{nc}(p,q), O_{nc}(n, \mathbf{C})$, $O_{nc}^*(2n)$, $Sp_{nc}(n, \mathbf{C})$, $Sp_{nc}(n, \mathbf{R})$, and $Sp_{nc}(p,q)$. Note that some of these are actually real or "real" pro-C^*-algebras, that is, inverse limits of the real or "real" C^*-algebras defined in Section 1 of [38]. Other dual groups can be constructed in a similar way. To give an idea of what kinds of algebras arise, we point out that in this context the "noncommutative torus" is $C^*(F_2)$, the full C^*-algebra of the free group on two generators.

A dual group A in the category of pro-C^*-algebras gives rise to an underlying group, by the procedure used to recover $U(n)$ from $U_{nc}(n)$. This group can actually be obtained more directly: it is the set of unital homomorphisms from A to $\mathbf{C}$, with operations obtained in a canonical manner from μ, ι, and χ. More generally, for every unital pro-C^*-algebra B, the set of unital homomorphisms from A to B has a natural group structure. The identity is $\varepsilon_B \circ \chi$, the product is $\varphi \cdot \psi = \delta_B \circ (\varphi * \psi) \circ \mu$, and the inverse of φ is $\varphi \circ \iota$. Here, ε_B and δ_B are as in Definition 2.4.1.

Voiculescu further defines dual actions of dual groups on (pro-)C^*-algebras, which in a similar way yield ordinary actions of the underlying groups on the same algebras. As an application, he defines a dual action of $U_{nc}(n, 1)$ on the Cuntz algebra $\mathcal{O}_n$ which gives rise in this manner to the action of $U(n, 1)$ on $\mathcal{O}_n$ defined in [70]. The construction of the dual action of $U_{nc}(n, 1)$ is more direct than the original construction of the action of $U(n, 1)$. For details of all of this, see [71].

2.5. Classifying spaces for the K-theory of C^*-algebras. In algebraic topology it is considered desirable to be able to represent the groups $h^n(X)$ of a generalized cohomology theory h^* in the form $h^n(X) \cong [X, E_n]$ for some appropriate classifying spaces E_n. Here square brackets denote homotopy classes of continuous maps, and the isomorphism is to be natural in X. The Brown Representation Theorem (see [14], [15], [1]) gives fairly general conditions under which this can be done for cohomology theories defined, for example, on the category of finite complexes. It should be noted, however, that the space E_n usually cannot be chosen to be a finite complex; rather, in general it will be infinite CW-complex.

The methods of [15] can, we believe, be used to prove a version of the Brown Representation Theorem for homotopy functors on small categories of C^*-algebras, such as the category of separable C^*-algebras. That is, the homotopy functor h can be represented as $h(A) \cong [E, A]$ for some appropriate pro-C^*-algebra E and all separable C^*-algebras A. Here square brackets denote homotopy classes of homomorphisms of pro-C^*-algebras. Unfortunately, the algebras E obtained this way will be inverse limits of very large inverse systems, and do not seem to be very useful. (The difficulty stems from the fact that we are still missing an analog of the category of finite complexes.) There are, however, very nice classifying σ-C^*-algebras for the functors K_0 and K_1, taken as being defined on the category of all C^*-algebras. (A good general reference for K-theory is [12].) We will describe these algebras in this section. Most of this material has not appeared elsewhere.

We should mention that Rosenberg has obtained in [62] a representation for K-theory which is analogous to what the topologists do for covariant functors. For example, he proves (Theorem 4.1) that $K_1(A) \cong [C_0(\mathbf{R}),\ A \otimes K]$, where K is the algebra of compact operators on a separable infinite-dimensional Hilbert space. Here, however, we imitate the procedure the topologists use for contravariant functors, but with all the arrows reversed. Since the functor $X \mapsto C(X)$ is contravariant, our results are more directly analogous to the topological results.

It is convenient to imitate the topologists in another respect as well: we work in the categories of pointed C^*-algebras and pointed pro-C^*-algebras. These are the noncommutative analogs of categories of topological spaces with base-points.

2.5.1. DEFINITION. A pointed (pro-)C^*-algebra is a pair (A, α), where A is a unital (pro-)C^*-algebra and $\alpha : A \to \mathbf{C}$ is a unital homomorphism.

A morphism from (A, α) to (B, β) is a homomorphism of (pro-)C^*-algebras $\varphi : A \to B$ such that $\beta \circ \varphi = \alpha$. We will usually drop α from the notation. We write $\mathrm{Hom}_+(A, B)$ for the set of morphisms of pointed (pro-)C^*-algebras, and we write $[A, B]_+$ for the set of homotopy classes in $\mathrm{Hom}_+(A, B)$.

We should note that the assignment $A \mapsto (A^+, \alpha)$, where $\alpha(a + \lambda \cdot 1) = \lambda$, is a category equivalence from (pro-)C^*-algebras to pointed (pro-)C^*-algebras. (To go the other way, note that if (A, α) is pointed, then $A \cong \mathrm{Ker}(\alpha)^+$.)

We construct the classifying algebra for K_1 first, because it is easier. Let $U_{nc}(n)$ be the noncommutative unitary group discussed in the previous section, and call its generators 1 and $x_{i,j,n}$ $(1 \le i, j \le n)$. By its definition, there is a unique homomorphism $\kappa_n : U_{nc}(n+1) \to U_{nc}(n)$ such that

$$\kappa_n(x_{i,j,n+1}) = \begin{cases} x_{i,j,n} & 1 \le i, j \le n \\ 1 & i = j = n+1 \\ 0 & \text{otherwise.} \end{cases}$$

Define $U_{nc}(\infty) = \varprojlim U_{nc}(n)$ with respect to the maps κ_n. Since the maps κ_n are homomorphisms of dual groups, $U_{nc}(\infty)$ is a dual group in the category of σ-C^*-algebras, as in the previous section.

We note that a dual group A is always a pointed algebra in a natural way, using the homomorphism $\chi : A \to \mathbf{C}$ of Definition 2.4.1. Furthermore, if B is any pointed (pro-)C^*-algebra, then $\mathrm{Hom}_+(A, B)$ is a subgroup of $\mathrm{Hom}(A, B)$ in the group structure defined at the end of the previous section, and $[A, B]_+$ is a quotient group of $\mathrm{Hom}_+(A, B)$. Therefore, for any (pro-)C^*-algebra B, we obtain a group $[U_{nc}(\infty), B^+]_+$.

2.5.2. PROPOSITION (Compare [54], Remark 4.12.) *For every C^*-algebra B, there is a natural isomorphism of groups $[U_{nc}(\infty), B^+]_+ \cong K_1(B)$.*

Proof. Since a homotopy of homomorphisms to B^+ is the same as a homomorphism to $C([0,1]) \otimes B^+$, it follows from Lemma 1.1.5 that $[U_{nc}(\infty), B^+]_+ \cong \varinjlim [U_{nc}(n), B^+]_+$. Now by the universal property defining $U_{nc}(n)$, the group

$[U_{nc}(n), B^+]_+$ is the group of path-components of the group of unitaries in $M_n(B^+)$ whose image in $M_n(\mathbf{C})$ under the obvious map is 1. Since the unitary group of $M_n(\mathbf{C})$ is connected, this is the same as the group of path-components $(U/U_0)(M_n(B^+))$ (notation as in Section 1.6). It is well known that

$$\varinjlim (U/U_0)(M_n(B^+)) \cong K_1(B),$$

and one easily checks that the maps in the direct limit agree with those in the limit $\varinjlim [U_{nc}(n), B^+]_+$. Q.E.D.

For K_0 the situation is trickier. We do not know of a nice dual group which will serve as the classifying algebra, so we need the following weaker notion.

2.5.3. DEFINITION. A *homotopy dual group* in the category of (pro-)C^*-algebras is a unital (pro-)C^*-algebra A equipped with pointed morphisms $\mu : A \to A *_{\mathbf{C}} A$, $\iota : A \to A$, and $\chi : A \to \mathbf{C}$ such that the relations of Definition 2.4.1 hold up to homotopy in the category of pointed (pro-)C^*-algebras. (Here A is pointed using χ, and $A *_{\mathbf{C}} A$ is pointed using $\chi * \chi$.) A is called abelian if, with $\varphi : A *_{\mathbf{C}} A \to A *_{\mathbf{C}} A$ being the obvious flip homomorphism, the maps $\varphi \circ \mu$ and μ are homotopic in the category of pointed (pro-)C^*-algebras.

In Proposition 2.5.2, the fact that $[U_{nc}(\infty), B^+]_+$ is always an abelian group is explained by the fact that $U_{nc}(\infty)$ is homotopy abelian, even though it is not abelian (that is, $\varphi \circ \mu \neq \mu$). We also note that if A is an abelian homotopy dual group, then $[A, B]_+$ still has a natural abelian group structure, because $\mathrm{Hom}_+(A, B)$ is at least still a group up to homotopy.

The classifying algebra that we construct for K_0 is going to be the unitization of the algebra $W_\infty(q\mathbf{C}) = \varprojlim W_n(q\mathbf{C})$, where $q\mathbf{C}$ is the classifying algebra for quasihomomorphisms from $\mathbf{C}$ defined by Cuntz in [21] and W_n is a left adjoint for the $n \times n$ matrix functor $B \mapsto M_n(B)$. We refer to [21] for the definition of qA for any C^*-algebra A; here we only say that its key property is that $[qA, B]$ can be canonically identified with the set of homotopy classes of quasihomomorphisms form A to B. The algebra $q\mathbf{C}$ should be though of as the universal C^*-algebra on the difference of two projections. It can be shown to be isomorphic to

$$\{f : [0,1] \to M_2(\mathbf{C}) : f(0) \text{ is diagonal and } f(1) = 0\}.$$

We now construct W_n.

2.5.4. DEFINITION. Let A be any C^*-algebra. Then $W_n(A)$ is defined to be the C^*-algebra on generators $x_n(a,i,j)$ for $a \in A$ and $1 \le i,j \le n$, subject to the relations that the $n \times n$ matrices $x_n(a) = (x_n(a,i,j))_{i,j=1}^n$ satisfy all the algebraic relations satisfied by the corresponding elements of A.

The relations used in this definition are admissible (Definition 1.3.2), because in any representation one has $\|x_n(a,i,j)\| \le \|a\|$. Therefore $W_n(A)$ is in fact a C^*-algebra. Its properties are given in the following proposition.

2.5.5. PROPOSITION. (1) *W_n is a functor from C^*-algebras to C^*-algebras.*

(2) *W_n is a left adjoint for M_n, that is, there are bijections*

$$\mathrm{Hom}(W_n(A), B) \to \mathrm{Hom}(A, M_n(B)) \quad \textit{and} \quad [W_n(A), B] \to [A, M_n(B)]$$

which are natural in A and B. They are given by assigning to $\varphi \in \mathrm{Hom}(W_n(A), B)$ the homomorphism $a \mapsto (\varphi(x_n(a,i,j)))$ from A to $M_n(B)$.

(3) *There are homomorphisms from $W_{n+1}(A)$ to $W_n(A)$ given by $x_{n+1}(a,i,j) \mapsto x_n(a,i,j)$ if $i,j \le n$ and $x_{n+1}(a,i,j) \mapsto 0$ otherwise, and they are a natural transformation of functors.*

(4) *Let A be a C^*-algebra, and let $\iota_0 : A \to A$ be an endomorphism such that ι_0^2 is homotopic to id_A and $a \mapsto \begin{pmatrix} \iota_0(a) & 0 \\ 0 & a \end{pmatrix}$ is homotopic to the zero map. Let $W_\infty(A) = \varprojlim W_n(A)$, using the maps in (3). Then $W_\infty(A)^+$ is an abelian homotopy dual group in the category of σ-C^*-algebras in a natural way.*

The first three parts of this proposition are immediate from the construction of $W_n(A)$. In the fourth part, μ, ι, and χ are given as follows. The map ι is the unitized inverse limit of the maps $W_n(\iota_0)$, and $\chi : W_\infty(A)^+ \to \mathbf{C}$ is the obvious homomorphism. To define μ, we write the generators of $W_\infty(A)$ as $x(a,i,j)$ for $a \in A$, $i,j \in \mathbf{N}$, the generators of the first factor of $W_\infty(A) * W_\infty(A)$ as $x'(a,i,j)$, and the generators of the second factor as $x''(a,i,j)$. The μ is the unitization of the map

$$x(a,i,j) \mapsto \begin{cases} 0 & i \neq j \bmod 2 \\ x'(a,i_0,j_0) & i = 2i_0 - 1,\ j = 2j_0 - 1 \\ x''(a,i_0,j_0) & i = 2i_0,\ j = 2j_0. \end{cases}$$

This map is an analog of the homomorphism

$$(A \otimes K) \oplus (A \otimes K) \xrightarrow{\varphi} A \otimes M_2(K) \xrightarrow{\psi} A \otimes K,$$

in which $\varphi(a,b) = \left(\begin{smallmatrix} a & 0 \\ 0 & b \end{smallmatrix}\right)$ and ψ comes from an isomorphism $M_2(K) \cong K$. Thus, the operation on homotopy classes of maps determined by μ is essentially the direct sum. The proof of this part of the proposition is a bit messy, and is omitted.

One can now use an argument similar to the proof of Proposition 2.5.2, together with the fact that all elements of $K_0(B)$ can be expressed as differences of projections in some $M_n(B^+)$, to prove:

2.5.6. PROPOSITION. *For any C^*-algebra B then is a natural isomorphism of groups $[W_\infty(q\mathbf{C})^+, B^+]_+ \cong K_0(B)$.*

We close this section with a number of remarks. First, in spite of this proposition and the properties of qA, it is not usually true that $[W_\infty(qA)^+, B^+]_+ \cong KK^0(A,B)$. (See [12] for the definition of KK-theory.) One must instead replace W_∞ by a left adjoint for the functor $K \otimes -$. There is one, but it takes values which are uncountable inverse limits of C^*-algebras, and is therefore much harder to deal with. Note, however, that when $C_0(\mathbf{R})$ is equipped with the endomorphism $\iota_0(f)(t) = f(-t)$, then there is an isomorphism of σ-C^*-algebras $W_\infty(C_0(\mathbf{R}))^+ \cong U_{nc}(\infty)$ which identifies the maps μ, ι, and χ up to homotopy. Thus, the classifying algebra for K_1 can also be obtained via Proposition 2.5.5. Incidentally, this also shows that $U_{nc}(\infty)$ is homotopy abelian.

Next, we look at the representable K-theory of our classifying algebras. The traditional classifying spaces for K-theory, namely $U = \varinjlim U(n)$ for K_1 and $\mathbf{Z} \times BU = \mathbf{Z} \times \varinjlim BU(n)$ for K_0, have fairly complicated K-theory. (See, for example, [69], 16.33.) However, it can be shown that the maps $W_{n+1}(A) \to W_n(A)$ are all isomorphisms on K-theory, from which it follows that, with RK_* as defined in Section 1.6, we have $RK_*(W_\infty(A)) \cong K_*(A)$. Therefore $W_\infty(q\mathbf{C})$ has $RK_0 = \mathbf{Z}$ and $RK_1 = 0$, while the kernel of the map $U_{nc}(\infty) \to \mathbf{C}$ has $RK_0 = 0$ and $RK_1 = \mathbf{Z}$. These groups couldn't be simpler.

We have so far only discussed the representability of K-theory for C^*-algebras, not the representability of RK_*. Although we have not quite been able to prove it, we believe that the following conjecture is true.

2.5.7. CONJECTURE. For each σ-C^*-algebra B there are natural isomorphisms $[W_\infty(q\mathbf{C})^+, B^+]_+ \cong RK_0(B)$ and $[U_{nc}(\infty), B^+]_+ \cong RK_1(B)$.

Finally, we mention that the relationships between K_0 and K_1 suggest certain relationships between their classifying algebras. To make this precise, however, requires the loop algebra of the next section.

2.6. **The noncommutative loop space.** In the first section of [62], Rosenberg writes that the first thing one would want in a good category containing both C^*-algebras and infinite CW-complexes is a left adjoint for the suspension functor. In this section, we construct such a left adjoint in the category of pointed pro-C^*-algebras. As in the previous section, this material has not appeared elsewhere.

2.6.1. DEFINITION. If (A, α) is a pointed pro-C^*-algebra, than its *suspension* is the pro-C^*-algebra

$$\Sigma A = \{f : S^1 \to A \text{ continuous} : \alpha(f(\varsigma)) \cdot 1 = f(1) \text{ for all } \varsigma \in S^1\},$$

together with the homomorphism $ev_1 : \Sigma A \to \mathbf{C}$ of evaluation at 1. Here S^1 is identified with $\{\varsigma \in \mathbf{C} : |\varsigma| = 1\}$. (Note that ev_1 makes sense as a homomorphism to $\mathbf{C}$, since $f(1) \in \mathbf{C} \cdot 1$ for $f \in \Sigma A$.)

Note that $\Sigma(A^+) \cong (SA)^+$, where SA is the conventional suspension $C_0(\mathbf{R}) \otimes A$. A left adjoint for Σ in the pointed category immediately gives a left adjoint for S in the category of pro-C^*-algebras and arbitrary homomoprhisms, simply by taking the kernel of the homomorphism to $\mathbf{C}$ which comes with the pointed pro-C^*-algebra. It is a left adjoint for S that Rosenberg really had in mind, since what he calls Σ we call S. However, our approach more closely matches the topologists' conventions.

2.6.2. DEFINITION. Let (A, α) be a pointed pro-C^*-algebra. We construct a pointed pro-C^*-algebra ΩA in terms of generators and relations as follows. Let the generating set G consist of the symbols $z(a, \varsigma)$ for $a \in A$ and $\varsigma \in S^1$, and let the relations R be as follows:

(1) The map $(a, \varsigma) \mapsto z(a, \varsigma)$ is continuous.

(2) For each fixed $\varsigma \in S^1$, the elements $z(a, \varsigma)$ satisfy all the algebraic relations satisfied by the corresponding elements of A.

(3) $z(1, \varsigma) = z(1, 1)$ for all $\varsigma \in S^1$.

(4) $z(a, 1) = z(\alpha(a)1, 1)$ for all $a \in A$.

Then set $\Omega A = C^*(G, R)$. The required homomorphism from ΩA to $\mathbf{C}$ is given by $z(a, \varsigma) \mapsto \alpha(a)$ for $a \in A$ and $\varsigma \in S^1$.

We note that the relations R are weakly admissible (see Example 1.3.5), so that ΩA is in fact a pro-C^*-algebra. If A is a pointed C^*-algebra, given as $C^*(G_0, R_0)$ with $1 \in G_0$, then the generators and relations can be simplified as follows. We use only the generators $z(a, \varsigma)$ with $a \in G_0$; in (1) we require that for each fixed $g \in G_0$, the map $\varsigma \mapsto z(g, \varsigma)$ be continuous; in (2) we need only the relations R_0, not all the relations holding in A; and in (4) we need only

$z(g,1) = z(\alpha(g)1,1)$ for $g \in G_0$. If R_0 is an admissible (Definition 1.3.2) set of algebraic and norm relations, we can then use the method of Example 1.3.9 to obtain ΩA as an inverse limit of C^*-algebras given by admissible sets of algebraic and norm relations, over the directed set D^{G_0}, where D is as in Example 1.3.9. If G_0 is finite, then one can get away with using D itself, since the image of the diagonal embedding of D in D^{G_0} is then cofinal. However, D has no countable cofinal subset, so ΩA is never a σ-C^*-algebra, except in trivial cases. This is unfortunate, because it make ΩA very difficult to work with.

2.6.3. THEOREM. *The map* $\Phi : \mathrm{Hom}_+(\Omega A, B) \to \mathrm{Hom}_+(A, \Sigma B)$, *defined by* $\Phi(\varphi)(a)(\varsigma) = \varphi(z(a,\varsigma))$, *is a natural bijection, and also defines a natural bijection* $[\Omega A, B]_+ \to [A, \Sigma B]_+$. *In particular,* Ω *is a left adjoint for the functor* Σ.

Proof. The statement about homotopy classes follows from the statement about homomorphisms on replacing B by $B \otimes C([0,1])$. Therefore we only consider the statement about homomorphisms. It is sufficient to prove that Φ defines a one to one correspondence between pointed homomorphisms from A to ΣB and representations ρ in B of (G,R) which correspond to pointed homomorphisms from ΩA to B. If $\beta : B \to \mathbf{C}$ is the homomorphism making B a pointed algebra, then the conditions on ρ are $\rho(z(1,1)) = 1$ and $\beta \circ \rho(z(a,\varsigma)) = \alpha(a)$ for $a \in A$ and $\varsigma \in S^1$. Using the universal property defining an inverse limit on the isomorphism $B \cong \varprojlim_{p \in S(\mathrm{Ker}(\beta))} (\mathrm{Ker}(\beta)_p)^+$, and conditions (3) and (4) of the definition of weak admissibility (Definition 1.3.4), we see that it suffices to consider the case in which B is a C^*-algebra.

Let $\rho : G \to B$ be a representation of (G,R) such that $\rho(z(1,1)) = 1$ and $\beta \circ \rho(z(a,\varsigma)) = \alpha(a)$. Define $\psi : A \to \Sigma B$ by $\psi(a)(\varsigma) = \rho(z(a,\varsigma))$. Then ψ is certainly a unital *-homomorphism satisfying $ev_1(\psi(a)) = \rho(z(a,1)) = \alpha(a)$, as desired. (Note that $\psi(a) \in \Sigma B$, by the relations (3) and (4).) That ψ is continuous follows from relation (1) and the compactness of S^1 : if $a_i \to a$ then the joint continuity of $\rho : A \times S^1 \to B$ forces $\rho(z(a_i, \varsigma))$ to converge uniformly in ς to $\rho(z(a, \varsigma))$. This shows that the definition of Φ makes sense.

Conversely, let $\psi : A \to \Sigma B$ be a pointed morphism. Define $\rho(z(a, \varsigma)) = \psi(a)(\varsigma)$. Clearly $\rho(z(1, 1)) = 1$ and $\beta \circ \rho(z(a, \varsigma)) = \beta(\psi(a)(\varsigma)) = \beta(\psi(a)(1)) = \alpha(a)$. Also, the relations of the definition all hold: (1) because ψ is continuous for the supremum norm on ΣB; (2) because evaluation at ς composed with ψ is a homomorphism for all ς; (3) because ψ is unital; and (4) because $\psi(a)(1) = \beta(\psi(a)(1)) \cdot 1$ for $a \in A$. Q.E.D.

We should note that if $A = C(X)$ for some compactly generated completely Hausdorff space X, and if $\alpha = ev_{x_0}$, evaluation at x_0, for some $x_0 \in X$, then the abelianization of ΩA (Definition 1.5.7) is $C(\Omega(X, x_0))$, where $\Omega(X, x_0)$ is the usual loop space of X relative to the basepoint x_0 ([69], 2.6), and is equipped with the compactly generated topology as discussed following I.4.18 in [74]. Note, however, that $\Omega(C(X))$ is essentially never commutative, because if $\varphi : C(X) \to \Sigma B$ is a homomorphism, then there is generally no reason for the ranges of the homomorphisms $ev_\varsigma \circ \varphi : C(X) \to B$ to commute with each other. On the other hand, ΩA can be trivial for noncommutative algebras A. For example, if A is the unitization of a simple C^*-algebra A_0, and if $\varphi : A \to \Sigma B$ is a homomorphism, then each homomoprhism $ev_\varsigma \circ \varphi|_{A_0}$ is homotopic to $ev_1 \circ \varphi|_{A_0} = 0$. Since A_0 is simple, this can only happen if $ev_\varsigma \circ \varphi|_{A_0} = 0$ for all ς, that is, $\varphi|_{A_0} = 0$. It follows that $\Omega A = \mathbf{C}$.

Loop spaces have many uses in topology. We will discuss one application here, namely the notion of an Ω-spectrum. A spectrum is a sequence of pointed spaces E_n for $n \in \mathbf{Z}$, usually required to be CW-complexes, with some relation between E_{n+1} and the suspension $\Sigma E_n = E_n \wedge S^1$. (This is the commutative version of the suspension of Definition 2.6.1.) Then one wants to define a cohomology theory E by $E^n(X) = [X, E_n]_+$ for, say, pointed CW-complexes X. In order for this theory to behave properly on suspensions, one wants to have ΩE_{n+1} weakly homotopy equivalent to E_n, in which case the spectrum is called an Ω-spectrum. (For this, and more details, see Section III.2 of [2]. Another approach can be found in Chapter 8 of [69].) Returning now to the classifying algebras for K-theory of the last section, this suggests the following conjecture:

2.6.4. CONJECTURE. These are homotopy equivalences

$$\Omega(W_\infty(q\mathbf{C})^+) \simeq U_{nc}(\infty) \quad \text{and} \quad \Omega U_{nc}(\infty) \simeq W_\infty(q\mathbf{C})^+$$

which determine the isomorphisms in K-theory $K_0(SB) \cong K_1(B)$ and $K_1(SB) \cong K_0(B)$ respectively.

This is the noncommutative version of the original Bott periodicity theorem, which asserts the existence of homotopy equivalences $\Omega(\mathbf{Z} \times BU) \simeq U$ and $\Omega U \simeq \mathbf{Z} \times BU$, where U is the infinite unitary group $\varinjlim U(n)$. (Compare [2], Example III.2.2.) This conjecture should imply Conjecture 2.5.7. It should also enable one to extend RK_* to the category of all pro-C^*-algebras in a reasonable way, via the definitions $RK_0(B) = [W_\infty(q\mathbf{C})^+, B^+]_+$ and $RK_1(B) = [U_{nc}(\infty), B^+]_+$. We note that the fact that these definitions are correct for C^*-algebras gives us a weak version of the homotopy equivalences of the conjecture. For example, there is an easily constructed homomorphism from $\Omega(W_\infty(q\mathbf{C})^+)$ to $U_{nc}(\infty)$ which induces an isomorphism $[\Omega(W_\infty(q\mathbf{C})^+), B^+]_+ \to [U_{nc}(\infty), B^+]_+$ for any C^*-algebra B.

Several other comments on noncommutative loop spaces are in order. First, there are many variants of the construction. For example, if one considers unpointed pro-C^*-algebras and drops relations (3) and (4) in Definition 2.6.2, then one obtains a noncommutative version of the free loop space. The noncommutative free loop space of $\mathbf{C}$ is already large, being in fact similar to the universal pro-C^*-algebra on a continuous path of projections of Example 1.3.3(6). Secondly, there was nothing special about the circle in the construction of ΩA, except for its potential usefulness in applications. The same methods give a left adjoint to the functor

$B \mapsto B \otimes C(X)$, or the pointed version of this functor, for any (pointed) compact Hausdorff space X.

REFERENCES

[1] J. F. Adams, *A variant of E.H. Brown's representation theorem*, Topology **10** (1971), 185-198.

[2] J. F. Adams, *Stable Homotopy and Generalized Homology*, Part III, Chicago Lectures in Math., University of Chicago Press, 1974.

[3] C. A. Akemann, G. K. Pedersen, and J. Tomiyama, *Multipliers of C^*-algebras*, J. Functional Anal. **13** (1973), 277-301.

[4] G. R. Allan, *On a class of locally convex algebras*, Proc. London Math. Soc. (3) **17** (1967), 91-114.

[5] R. Arens, *A topology for spaces of transformations*, Ann. of Math. **47** (1946), 480-495.

[6] R. Arens, *A generalization of normed rings*, Pacific J. Math **2** (1952), 455-471.

[7] W. Arveson, *The harmonic analysis of automorphism groups*, Proc. Symposia Pure Math. vol. 38, part 1, 199-269, Amer. Math. Soc., 1982.

[8] M. F. Atiyah and I. G. Macdonald, *Introduction to Commutative Algebra*, Addison-Wesley, Reading, Mass., 1969.

[9] M. F. Atiyah and G. B. Segal, *Equivariant K-theory and completion*, J. Differential Geometry **3** (1969), 1-18.

[10] D. Avitzour, *Free products of C^*-algebras*, Trans. Amer. Math. Soc. **271** (1982), 423-435.

[11] B. Blackadar, *Shape theory for C^*-algebras*, Math. Scand. **56** (1985), 249-275.

[12] B. Blackadar, *K-Theory for Operator Algebras*, MSRI publications no. 5, Springer-Verlag, New York, Berlin, Heidelberg, London, Paris, Tokyo, 1986.

[13] R. M. Brooks, *On representing F^*-algebras*, Pacific J. Math. **39** (1971), 51-69.

[14] E. H. Brown, *Cohomology theories*, Ann. of Math (2) **75** (1962), 467-484 and **78** (1963), 201.

[15] E. H. Brown, *Abstract homotopy theory*, Trans. Amer. Math. Soc. **119** (1965), 79-85.

[16] L. G. Brown, *Ext of certain free product C^*-algebras*, J. Operator Theory **6** (1981), 135-141.

[17] L. G. Brown, R. G. Douglas, and P. A. Fillmore, *Extensions of C^*-algebras and K-homology*, Ann. of Math. (2) **105** (1977), 265-324.

[18] A. Connes, *An analogue of the Thom isomorphism for crossed products of a C^*-algebra by an action of* **R**, Advances in Math. **39** (1981), 31-55.

[19] J. Cuntz, *Simple C^*-algebras generated by isometries*, Commun. Math. Phys. **57** (1977), 173-185.

[20] J. Cuntz, *K-theoretic amenability for discrete groups*, J. Reine Ang. Math. **344** (1983), 180-195.

[21] J. Cuntz, *A new look at KK-theory*, *K*-Theory **1** (1987), 31-51.

[22] J. Cuntz, G. A. Elliott, F. M. Goodman, and P.E.T. Jorgensen, *On the classification of noncommutative tori*, II, C.R. Math. Rep. Acad. Sci. Canada **7** (1985), 189-194.

[23] J. Cuntz and W. Krieger, *A class of C^*-algebras and topological Markov chains*, Invent. Math. **56** (1980), 251-268.

[24] H. G. Dales, *Automatic continuity: a survey*, Bull. London Math. Soc. **10** (1978), 129-183.

[25] S. Disney, G. A. Elliott, A. Kumjian, and I. Raeburn, *On the classification of noncommutative tori*, C.R. Math. Rep. Acad. Sci. Canada **7** (1985), 137-141.

[26] P. G. Dixon, *Generalized B^*-algebras*, Proc. London Math. Soc. (3) **21** (1970), 693-715.

[27] A. Dold, *Partitions of unity in the theory of fibrations*, Ann. of Math. **78** (1963), 223-255.

[28] J. Dugundji, *Topology*, Allyn and Bacon, Boston, 1966.

[29] G. A. Elliott, *On the K-theory of the C^*-algebra generated by a projective representation of a torsion-free discrete abelian group*, in: *Operator Algebras and Group Representations* vol. 1, 157-184, Pitman, London, 1984.

[30] M. Fragoulopoulou, *Spaces of representations and enveloping l.m.c. *-algebras*, Pacific J. Math. **95** (1981), 61-73.

[31] M. Fragoulopoulou, *Representations of tensor product l.m.c. *-algebras*, Manuscr. Math. **42** (1983), 115-145.

[32] M. Fragoulopoulou, *Kadison's transitivity for locally C^*-algebras*, J. Math. Anal. Appl. **108** (1985), 422-429.

[33] M. Fragoulopoulou, *Symmetric topological *-algebras*, Abstracts Amer. Math. Soc. **6** (1985) 415.

[34] M. Fritsche, *On the existence of dense ideals in LMC*-algebras*, Z. Anal. Anw. **1** (1982), 3, 81-84.

[35] M. Fritsche, *Über die Struktur Maximaler Ideale in LMC*-Algebren*, Z. Anal.

Anw. **4** (1985), 3, 201-205.

[36] A. Inoue, *Locally C^*-algebra*, Mem. Fac. Sci. Kyushu Univ. **25** (1971), 197-235.

[37] P. Julg, *K-theorie equivariante et produits croisés*, C. R. Acad. Sci. Paris Ser. I, **292** (1981), 629-632.

[38] G. G. Kasparov, *Hilbert C^*-modules: theorems of Stinespring and Voiculescu*, J. Operator Theory **4** (1980), 133-150.

[39] G. G. Kasparov, *K-theory, group C^*-algebras, and higher signatures* (conspectus), parts 1 and 2, preprint Chernogolovka.

[40] G. Köthe, *Topologische Lineare Räume I* (Grundlehren der mathematischen Wissenschaften in Einzeldarstellungen No. 107), Springer-Verlag, Berlin, Göttingen, Heidelberg, 1960.

[41] A. J. Lazar and D. C. Taylor, *Multipliers of Pedersen's ideal*, Mem. Amer. Math. Soc. **169** (1976).

[42] J. Mack, S. A. Morris, and E. T. Ordman, *Free topological groups and the projective dimension of a locally compact abelian group*, Proc. Amer. Math. Soc. **40** (1972), 303-308.

[43] A. Mallios, *Vector bundles and K-theory over topological algebras*, J. Math. Anal. Appl. **92** (1983), 452-506.

[44] A. Mallios, *Hermitian K-theory over topological *-algebras*, J. Math. Anal. Appl. **106** (1985), 454-539.

[45] A. Mallios, *Continuous vector bundles on topological algebras*, J. Math. Anal. Appl. **113** (1986), 245-254.

[46] E. A. Michael, *Locally multiplicatively-convex topological algebras*, Mem. Amer. Math. Soc. **11** (1952).

[47] J. W. Milnor, *Construction of universal bundles II*, Ann. of Math **63** (1956), 430-436.

[48] J. W. Milnor, *Axiomatic homology theory*, Pacific J. Math **12** (1962), 337-341.

[49] J. A. Mingo, *K-Theory and multipliers of stable C^*-algebras*, Trans. Amer. Math. Soc. **299** (1987), 397-411.

[50] G. K. Pedersen, *Measure theory for C^*-algebras*, Math. Scand. **19** (1966), 131-145.

[51] G. K. Pedersen, *C^*-Algebras and their Automorphism Groups*, Academic Press, London, New York, San Francisco, 1979.

[52] N. C. Phillips, *Equivariant K-Theory and Freeness of Group Actions on C^*-Algebras*, Springer Lecture Notes in Math. no. 1274, Springer-Verlag, Berlin, Heidelberg, New York, London, Paris, Tokyo, 1987.

[53] N. C. Phillips, *Inverse limits of C^*-algebras*, J. Operator Theory, to appear.

[54] N. C. Phillips, *Representable K-theory for σ-C^*-algebras*, preprint.

[55] N. C. Phillips, *The Atiyah-Segal completion theorem for C^*-algebras*, preprint.

[56] N. C. Phillips, *A new approach to the multipliers of Pedersen's ideal*, Proc. Amer. Math. Soc., to appear.

[57] M. Pimsner and D. Voiculescu, *Exact sequences for K-groups and Ext-groups of certain crossed product C^*-algebras*, J. Operator Theory **4** (1980), 93-118.

[58] M. Pimsner and D. Voiculescu, *K-groups of reduced crossed products by free groups*, J. Operator Theory **8** (1982), 131-156.

[59] M. A. Rieffel, *Actions of finite groups on C^*-algebras*, Math. Scand. **47** (1980), 157-176.

[60] M. A. Rieffel, *C^*-algebras associated with irrational rotations*, Pacific J. Math. **93** (1981), 415-429.

[61] M. A. Rieffel, *Projective modules over higher dimensional noncommutative tori*, Canadian Math. J., to appear.

[62] J. Rosenberg, *The role of K-theory in noncommutative algebraic topology*, in: *Operator Algebras and K-Theory* (Contemporary Math. vol. 10), 155-182, Amer. Math. Soc., 1982.

[63] K. Schmüdgen, *Über LMC*-Algebren*, Math. Nachr. **68** (1975), 167-182.

[64] C. Schochet, *Topological methods for C^*-algebras III: axiomatic homology*, Pacific J. Math. **114** (1984), 399-445.

[65] G. Segal, *The representation ring of a compact Lie group*, Publ. Math. Inst. Hautes Etudes Sci. **34** (1968), 113-128.

[66] G. Segal, *Equivariant K-theory*, Publ. Math. Inst. Hautes Etudes Sci. **34** (1968), 129-151.

[67] E. Spanier, *Quasitopologies*, Duke Math. J. **30** (1963), 1-14.

[68] M. Spivak, *A Comprehensive Introduction to Differential Geometry*, Vol. 1 (2nd ed.) Publish or Perish, Houston, 1979.

[69] R. M. Switzer, *Algebraic Topology - Homotopy and Homology*, Springer-Verlag, Berlin, Heidelberg, New York, 1975.

[70] D. Voiculescu, *Symmetries of some reduced free product C^*-algebras*, in: *Operator Algebras and their Connection with Topology and Ergodic Theory*, Springer Lecture Notes in Math. no. 1132, 556-588, Springer-Verlag, Berlin, Heidelberg, New York, Tokyo, 1985.

[71] D. Voiculescu, *Dual algebraic structures on operator algebras related to free products*, J. Operator Theory **17** (1987), 85-98.

[72] J. Weidner, *Topological Invariants for Generalized Operator Algebras*, Ph.D. Thesis, Heidelberg, 1987.

[73] C. Wenjen, *On seminormed *-algebras*, Pacific J. Math. **8** (1958), 177-186.

[74] G. W. Whitehead, *Elements of Homotopy Theory* (Graduate Texts in Mathematics No. 61), Springer-Verlag, New York, Heidelberg, Berlin, 1978.

[75] Xia Daoxing [Sia Do-Shin], *On polynormed algebras with involution*, Izv. Akad. Nauk. SSSR, Cer. Mat. **23** (1959), 509-528 (in Russian).

Partitioning non-compact manifolds and the dual Toeplitz problem

John Roe
Mathematical Institute
24-29 St. Giles'
Oxford.

Introduction

In [13], I introduced the idea of considering the index of an elliptic operator on a non-compact manifold as an element of the K-theory of a certain operator algebra, called there the algebra of "uniform operators". The results of [13] showed that in certain circumstances a trace could be constructed on this operator algebra, giving rise to a real-valued dimension function on its K-theory; the resulting real-valued index was then calculated by the heat equation method. This approach had already been employed in a different geometrical context by Connes [5].

In this paper I address a special case of the following general question: are there cyclic cocycles of dimension > 0 on the algebra of uniform operators? How do such cocycles relate to the geometry of the non-compact manifold, and what index theorems do they lead to? Notice that a trace is a cyclic cocycle of dimension zero. To investigate this problem, it seems that one might follow the following procedure. First, let the manifold be a non-compact homogeneous space, the algebra one of translation-invariant operators. This algebra can be analysed using representation theory. Second, find 'geometrical' representatives for the cyclic cohomology of this algebra. (Cf. Connes [6], Ch.I. §9). Third, extend these geometrical representatives to cocycles on the algebra of uniform operators over a manifold geometrically similar to the original homogeneous space.

About the simplest case of this procedure is studied here. Our exemplar is the one-dimensional manifold $\mathbb{R}$, and the relevant geometrical fact about it is that it can be partitioned into two pieces $\mathbb{R}^+$ and $\mathbb{R}^-$ meeting on a codimension one submanifold (the origin 0). Fourier analysis identifies the translation-invariant smoothing operators on $\mathbb{R}$ with mutliplication operators on $\hat{\mathbb{R}}$, the dual group of $\mathbb{R}$, and the geometrical decomposition of $\mathbb{R}$ leads to the analytical decomposition of $L^2(\hat{\mathbb{R}})$ into positive and negative Hardy spaces. The study of the interaction of multiplication operators with this decomposition is exactly the classical theory of Wiener-Hopf operators.

We will consider odd-dimensional manifolds M partitioned into two pieces by a compact hypersurface N. In this situation it turns out that (following [8]) we are able to define a 1-dimensional cyclic cohomology class for the algebra of uniform operators over M. An elliptic operator on M has an odd index in the K_1-group of this algebra, so that by pairing with the cyclic class we get a number. This number turns out to be the solution of an index problem on N, so can be evaluat by the classical index theorem. We obtain therefore a new index theorem on noncompact manifolds. As a corollary, we get results both old and new about complete metrics of positive scalar curvature.

The title of this paper arises from the following consideration. To get index theorems on odd-dimensional manifolds, one often proceeds by dividing the spectrum of a self-adjoint operator into two halves, thus getting a class in the odd K-homology of the manifold. This is then paired with certain K^1-classes (invertible matrix-valued functions on the manifold) to get a number. Cf. Baum and Douglas [2] for details of this, which can be thought of as a generalized Toeplitz operator construction. Our approach is dual: we divide the manifold, and pair with functions on the spectrum i.e. kernels representing functions of our elliptic operator. Is this related to Kasparov's [11] "Dirac" and "dual Dirac" constructions?

Acknowledgements

The ideas for this paper were worked out on a visit to the United States in April, 1986 and I should like to thank R. Douglas, S. Hurder and J. Kaminker for their hospitality. The motivating question was asked by M. Wodzicki, who pointed out the similarity between the algebras considered in [13] and those whose cohomology was calculated in [8].

§1. Non compact manifolds and partitions

Let M be a complete Riemannian manifold. We will also assume that M is oriented, so that it has a canonical volume form. We will use this volume form to identify functions and densities on M with distributions.

Let S be a hermitian vector bundle on M. By an operator on S we will always mean a continuous linear operators $A : C_c^\infty(S) \to C^\infty(S)$, which also maps $\mathcal{E}'(S)$ (the space of compactly supported distributional sections of S) continuously to $\mathcal{D}'(s)$ (the space of all distributional sections of S). The kernel theorem (Schwartz [14]) says that such an operator is represented by a distributional kernel: a distribution k_A on $M \times M$, with value $k_A(x,y)$ in $\hom(S_y, S_x)$, such that

$$As(x) = \int k_A(x,y)s(y)dy .$$

(1.1) Definition. Let $\mathcal{V}$ (or $\mathcal{V}(S)$) denote the algebra of operators A on S satisfying the following two properties:

a) A extends to a bounded operator on the Hilbert space $L^2(S)$.

b) The distributional kernel k_A of A is supported within a finite distance of the diagonal in $M \times M$.

Since

$$k_{AB}(x,z) = \int k_A(x,y)k_B(y,z)dy ,$$

one readily checks that $\mathcal{V}$ is an algebra. The operators A in $\mathcal{V}$ are properly supported; they preserve the property of having compact supports.

From property a), $\mathcal{V}$ is a subalgebra of the C*-algebra of bounded operators on $L^2(S)$. Let $\bar{\mathcal{V}}$ denote the closure of $\mathcal{V}$ in this C*-algebra.

$\mathcal{V}$ is closely related to the algebra of "uniform operators of order zero", introduced in [13]. However, no requirement of bounded geometry is imposed here.

(1.2) <u>Definition</u>. Let $\mathcal{X}$ denote the subalgebra of $\mathcal{V}$ consisting of operators A satisfying in addition the condition

c) The kernel k_A is a C^∞ function on $M \times M$ or equivalently, by a well-known result.

c') The operator A maps $\mathcal{D}'(S)$ to $C^\infty(S)$.

Then $\mathcal{X}$ is an ideal in $\mathcal{V}$. We define similarly $\bar{\mathcal{X}}$ to be the closure of $\mathcal{X}$ in the C* topology of $\bar{\mathcal{V}}$; it is an ideal in $\bar{\mathcal{V}}$, as is easily checked.

We shall now introduce the geometric objects which will give rise to cyclic cohomology classes on $\mathcal{X}$.

(1.3) <u>Definition</u>. A <u>partition of M</u> is a decomposition $M = M^+ \cup M^-$, where M^+ and M^- are open subsets of M having as common boundary a compact hypersurface N.

The normal bundle to N is clearly trivial, and it can be given an orientation by demanding that the positive normal direction should point from M^- to M^+. This orientation of the normal bundle and the given orientation of M then combine to fix an orientation of N. Usually, the letter E will denote a partition of M.

We shall now use an idea of Feigin and Tsygan [8] to define a cyclic cohomology class for the algebra X depending on a partition E of M. We need a simple lemma:

(1.4) <u>Lemma</u>. Let $k(x,y) \in \hom(S_y, S_x)$ be the kernel of an operator A on S. Suppose that k is compactly supported on $M \times M$, and that it is the restriction to some subset $L_1 \times L_2$ of a smooth function on $M \times M$. Then A is of trace class.

<u>Proof</u>: Let k be the restriction of the smooth kernel $\tilde{k}$ to $L_1 \times L_2$. Choose a smooth bump-function ϕ of compact support on $M \times M$ with $\phi = 1$ on $L_1 \times L_2$. Then $\phi\tilde{k}$ is a compactly supported smoothing kernel, so by a well-known result it is the kernel of a trace-class operator B. But now $A = P_1 B P_2$, where P_i is the operator of multiplication by the characteristic function of L_i. Since the trace-class operators form an ideal, the result follows. □

Now suppose given a partition $E = (M^+, M^-, N)$ of M, and let π denote the operator on S given by mutliplication by the characteristic function of M^+.

(1.5) <u>Lemma</u>: If $A \in X$, then $\pi A(1 - \pi)$ and $(1 - \pi)A\pi$ are operators of trace class.

Proof: Let k_A denote the smoothing kernel of A. Then the kernel of $\pi A(1 - \pi)$ is the restriction of k_A to $M^+ \times M^- \subset M \times M$. Since k_A has support within some distance R of the diagonal in $M \times M$, the kernel of $\pi A(1 - \pi)$ has support within distance $2R$ of $N \times N \subset M \times M$. Since M is complete, a $2R$-neighbourhood of the compact set $N \times N$ is relatively compact. Now we may apply lemma (1.4). □

(1.6) Proposition: Let $A, B \in \mathcal{X}$. Then the operator

$$[\pi A \pi, \pi B \pi] - \pi[A,B]\pi$$

is of trace class. (This operator is in some sense the "curvature" of π on $\mathcal{X}$). Moreover, the functional

$$\zeta_E : (A,B) \to \mathrm{Tr}([\pi A \pi, \pi B \pi] - \pi[A,B]\pi)$$

is a cyclic 1-cocycle on $\mathcal{X}$.

Proof: Notice that

$$[\pi A \pi, \pi B \pi] - \pi[A,B]\pi = -\pi A(1 - \pi)B\pi + \pi B(1 - \pi)A\pi$$

which is of trace class by (1.5). Now let's check that ζ_E is a cyclic cocycle. Clearly $\zeta_E(A,B) + \zeta_E(B,A) = 0$. We must check that $b\zeta_E = 0$. We compute

$$\begin{aligned} b\zeta_E(A,B,C) &= \zeta_E(AB,C) - \zeta_E(A,BC) + \zeta_E(CA,B) \\ &= \mathrm{Tr}[\pi\{-AB(1 - \pi)C + C(1 - \pi)AB \\ &\qquad +A(1 - \pi)BC - BC(1 - \pi)A \\ &\qquad -CA(1 - \pi)B + B(1 - \pi)CA\}\pi] \quad . \end{aligned}$$

Each of the six terms in this sum is of trace class. Now let us write

$$\mathrm{Tr}(\pi B(1-\pi)CA\pi) - \mathrm{Tr}(\pi AB(1-\pi)C\pi)$$

$$= \mathrm{Tr}(\pi B(1-\pi)C(1-\pi)A\pi) - \mathrm{Tr}(\pi A(1-\pi)B(1-\pi)C\pi).$$

This identity holds since

$$\mathrm{Tr}(\pi B(1-\pi)C\pi A\pi) = \mathrm{Tr}(\pi A\pi B(1-\pi)C\pi)$$

by the commutator property of the trace. Therefore, the formula for $b\zeta_E(A,B,C)$ becomes, on applying the analogous identity to three pairs of terms,

$$\begin{aligned} b\zeta_E(A,B,C) = \mathrm{Tr}[&-\pi A(1-\pi)B(1-\pi)C\pi + \pi C(1-\pi)A(1-\pi)B\pi \\ &+ \pi A(1-\pi)B(1-\pi)C\pi - \pi B(1-\pi)C(1-\pi)A\pi \\ &- \pi C(1-\pi)A(1-\pi)B\pi + \pi B(1-\pi)C(1-\pi)A\pi] \\ = 0,\ &\text{as required.} \end{aligned}$$

□

There is a natural notion of bordism for partitions: two partitions of M are <u>bordant</u> if the corresponding hypersurfaces are oriented bordant in M in the usual sense, (Atiyah [1]). Bordism can be defined equivalently to be the smallest equivalence relation compatible with the following assertion: let $E_1 = (M_1^+, M_1^-, N_1)$ and $E_2 = (M_2^+, M_2^-, N_2)$ be partitions, and suppose that $M_2^+ \subset M_1^+$, $M_2^- \cap M_1^+$ relatively compact; then E_1 and E_2 are bordant.

(1.7) <u>Proposition</u>: The cyclic cohomology class of ζ_E depends only on the bordism class of the partition E.

Proof: Let E_1 and E_2 be bordant; without loss of generality we may assume that they are related as in the assertion above. So, if π_1 and π_2 are the corresponding multiplication operators, we have $\pi_1 = \pi_2 + \phi$, ϕ being the operator of multiplication by the characteristic function of the relatively compact set $M_2^- \cap M_1^+$. Therefore

$$\zeta_{E_1}(A,B) - \zeta_{E_2}(A,B)$$

$$= \mathrm{Tr}\{-\pi_1 A(1-\pi_1)B\pi_1 + \pi_1 B(1-\pi_1)A\pi_1 + \pi_2 A(1-\pi_2)B\pi_2 - \pi_2 B(1-\pi_2)A\pi_2\}$$

$$= \mathrm{Tr}\{-\pi_1 A(1-\pi_1)B + \pi_1 B(1-\pi_1)A + \pi_2 A(1-\pi_2)B - \pi_2 B(1-\pi_2)A\} .$$

(We used the commutator property of the trace and the fact that π is a projection to write $\mathrm{Tr}(\pi A(1-\pi)B) = \mathrm{Tr}(\pi\cdot\pi A(1-\pi)B) = \mathrm{Tr}(\pi A(1-\pi)B\pi)$.)

Now let us substitute $\pi_1 = \pi_2 + \phi$;

$$\begin{aligned}\pi_1 A(1-\pi_1)B &= (\pi_2 + \phi)A(1 - \pi_2 - \phi)B \\ &= \pi_2 A(1-\pi_2)B + \phi A(1-\pi_2)B - \pi_2 A\phi B - \phi A\phi B .\end{aligned}$$

Hence

$$\begin{aligned}\zeta_{E_1}(A,B) - \zeta_{E_2}(A,B) = \mathrm{Tr}\{&(\phi B(1-\pi_2)A - \phi A(1-\pi_2)B) \\ &+ (\pi_2 A\phi B - \pi_2 B\phi A) \\ &+ (\phi A\phi B - \phi B\phi A)\} .\end{aligned}$$

By Lemma (1.4), the operators $A\phi$, $B\phi$, ϕA, ϕB, are all of trace class. Therefore $\mathrm{Tr}(\phi A\phi B) = \mathrm{Tr}(\phi B\phi A)$. Moreover

$$\mathrm{Tr}(\phi B(1-\pi_2)A) = \mathrm{Tr}((1-\pi_2)A\phi B) .$$

so

$$\begin{aligned}\zeta_{E_1}(A,B) - \zeta_{E_2}(A,B) &= \mathrm{Tr}(A\phi B - B\phi A) \\ &= \mathrm{Tr}(\phi BA) - \mathrm{Tr}(\phi AB) \\ &= b\sigma(A,B)\end{aligned}$$

where σ is the 0-cochain on X defined by

$$\sigma(X) = -\mathrm{Tr}(\phi X). \qquad \square$$

We now aim to show that there is a Banach algebra $\bar{X} \supset X$ to which ζ extends by continuity. Then by Connes [6], Ch. II Prop. 15, ζ_E will define a map (depending only on the bordism class of E) from $K_1(\bar{X})$ to $\mathbb{C}$. To define the algebra $\bar{X}$ we make use of the following result of Connes [7] :

(1.8) Proposition: Let B be a Banach algebra, $U \subset B$ a dense subalgebra, ζ a cyclic 1-cocycle on U. Suppose that for each $x \in U$, the functional

$$y \to \zeta(x,y)$$

extends to a continuous linear functional on B. Then there is an algebra B_1, $U \subseteq B_1 \subseteq B$, such that:

i) B_1 is a Banach algebra (in some norm);

ii) $i : B_1 \to B$ induces an isomorphism on K-theory;

iii) ζ extends by continuity to a cyclic cocycle on B_1.

Proof: (Taken from Connes' lecture notes). The continuity condition implies that there is a map $\delta : U \to B^*$ s.t. $\langle \delta(x), u \rangle = \zeta(x,y)$ if $x, y \in U$. To say that ζ is a cyclic cocycle translates to say that δ is a skew-symmetric derivation with values in the B-bimodule B^*; that is,

$$\delta(xy) = x\delta(y) + \delta(x)y$$

$$\langle \delta(x), y \rangle + \langle x, \delta(y) \rangle = 0 .$$

Now δ, considered as an unbounded operator, is closable; for if $x_n \to 0$, $\delta(x_n) \to \phi$, then for all $y \in u$

$$\langle \phi, y \rangle = \lim \langle \delta(x_n), y \rangle$$

$$= -\lim \langle x_n, \delta(y) \rangle = 0 .$$

Since U is dense, $\phi = 0$.

Let $\bar{\delta}$ be the closure of δ, and let $B_1 = \mathrm{dom}(\bar{\delta})$, equipped with the graph norm $|||x||| = ||x|| + ||\bar{\delta}x||$. Then B_1 is a Banach space, and in fact a Banach algebra since δ is a derivation. By continuity, $\bar{\delta}$ is a B^*-valued skew derivation on B_1, and so gives a cyclic cocycle on B_1.

It remains to show that $i : B_1 \to B$ induces an isomorphism on K-theory. By the density theorem ([6], Ch.I Appendix 3; [10]) it is enough to prove the following assertion: let $M_n(B_1^+)$, $M_n(B^+)$ be the $n \times n$ matrix algebras over the algebras B_1, B with adjoined unit; then if $a \in M_n(B_1^+)$ is invertible in $M_n(B^+)$, it is invertible in $M_n(B_1^+)$. To prove this, notice first that since there exist elements of $M_n(B_1^+)$ arbitrarily close to a^{-1}, it is enough to consider the case $a = 1 - x$, $||x|| = C < 1$. Then

$$a^{-1} = 1 + x + x^2 + \dots ;$$

we must show that if $x \in M_n(B_1^+)$, then this series converges in $M_n(B_1^+)$. We have

$$\begin{aligned} |||x^n||| &= ||x^n|| + ||\bar{\delta}(x^n)|| \\ &\le ||x||^n + ||\bar{\delta}(x)x^{n-1}|| + \dots + ||x^{n-1}\bar{\delta}(x)|| \\ &\le C^n + nC^{n-1}||\bar{\delta}(x)|| . \end{aligned}$$

Thus $\sum_{n=0}^{\infty} |||x^n||| < \infty$, and the result follows. □

Applying this result of Connes to our situation, $U = X$, $B = \bar{X}$, $\zeta = \zeta_E$, we find that we must check that for each fixed $A \in X$, the functional

$$B \to \zeta_E(A,B)$$

should extend continuously to $\bar{X}$. This is straightforward, however, since

$$\begin{aligned} |\zeta_E(A,B)| &= |\mathrm{Tr}(-\pi A(1-\pi)B + B(1-\pi)A\pi)| \\ &\le (||\pi A(1-\pi)||_1 + ||(1-\pi)A\pi||_1)||B|| \end{aligned}$$

where $||..||_1$ denotes the trace norm. We therefore find that there is a Banach algebra $\bar{\bar{X}}$, having the same K-theory as the C*-algebra $\bar{X}$, to which ζ_E extends by continuity. Thus ζ_E defines a map

$$\zeta_E^* : K_1(\bar{X}) \to \mathbb{C} .$$

§2. The odd index on a non-compact manifold

In this paragraph we shall define a new index invariant for certain self-adjoint elliptic operators on a non-compact complete Riemannian manifold M. This index invariant takes values in the group $K_1(\bar{X})$ constructed above. It vanishes for compact manifolds; but we shall see that it does not necessarily vanish for non-compact manifolds, by using the cyclic cohomology classes arising from partitions to detect it.

Let M and S, then, be as in §1, and let D be a first order formally self-adjoint elliptic operator on S. We require that D have bounded propagation speed; that is, that the supremum of the norm of the principal symbol of D over the unit cosphere bundle of M should be finite. Then we have

(2.1) Lemma: a) D is essentially self-adjoint.

b) There is a constant $c > 0$ such that for any $s \in C_c^\infty(S)$, the support of $e^{itD}s$ is contained within a $c|t|$-neighbourhood of the support of s. □

For proof of these facts, see Chernoff [4]. Most examples of such D that arise in geometry are generalized Dirac operators ([9], [13]).

We begin with a functional calculus lemma that gives criteria for operators $f(D)$ to lie in the algebras X, $\bar{X}$, V, and $\bar{V}$. Recall (e.g. from Taylor [15]) that the zero'th symbol class $S^0(\mathbb{R})$ is the collection of all C^∞ functions f on $\mathbb{R}$ such that for each k,

$$\sup\{|f^{(k)}(x)(1 + |x|)^k| : x \in \mathbb{R}\} < \infty .$$

In particular, functions $f \in S^0(\mathbb{R})$ are tempered distributions on $\mathbb{R}$, so their Fourier transforms are also tempered distributions.

(2.2) Proposition: Let $f \in S^0(\mathbb{R})$, with compactly supported Fourier transform. Then $f(D) \in \mathcal{U}$.

Proof: Since f is bounded, the spectral theorem shows that $f(D)$ is a bounded operator on $L^2(S)$. We must check that it is given by a distributional kernel supported within a bounded neighbourhood of the diagonal.

Let $s \in C_c(S)$. Then for each k, $D^k s \in L^2(S)$ and hence $D^k f(D)s = f(D)D^k s \in L^2(S)$. From the (local) elliptic estimates for D, it follows that $f(D)s$ is smooth; so $f(D)$ maps $C_c^\infty(S)$ to $C^\infty(S)$. Since $f(D)$ is the adjoint of the operator $\bar{f}(D)$, which also maps $C_c^\infty(S)$ to $C^\infty(S)$, we see that $f(D)$ maps $\mathcal{E}'(S)$ to $\mathcal{D}'(S)$. By the Schwartz kernels theorem, then, $f(D)$ is given by a distributional kernel.

To check that this kernel is supported within a bounded neighbourhood of the diagonal, it is enough to check that there is a number $R > 0$ such that for any compactly supported smooth section s, the support of $f(D)s$ is contained within an R-neighbourhood of the support of s. Choose R such that $\mathrm{supp}(\hat{f}) \subset \left[-\frac{R}{c}, \frac{R}{c}\right]$, c being the propagation speed. Now the function

$$t \to e^{itD}s$$

is a smooth function with values in the Frechet space $C^\infty(S)$. We may therefore write

$$f(D)s = \frac{1}{2\pi}\int \hat{f}(t)e^{itD}s\,dt \qquad (*)$$

where $\hat{f}(t)$ is a compactly supported distribution. Then the support of $e^{itD}s$ is contained within an R-neighbourhood of the support of s for all $t \in \mathrm{supp}(\hat{f})$, and the desired result follows. □

Remark: We will need to notice that formula (*) in fact proves also that $f(D)s$ depends only on the geometry of M in an R-neighbourhood of $\mathrm{supp}(s)$, since $s_t = e^{itD}s$ is the unique solution of $\dot{s}_t = iDs_t$, $s_0 = s$, in such a neighbourhood

(2.3) Proposition:

i) If $f \in S^0(\mathbb{R})$, then $f(D) \in \overline{V}$;

ii) If $\hat{f} \in C_c^\infty(\hat{\mathbb{R}})$, then $f(D) \in X$;

iii) If $f \in C_0(\mathbb{R})$, then $f(D) \in \overline{X}$.

Proof: i) Let $f \in S^0(\mathbb{R})$. Then all the derivatives of f of order at least 2 are integrable, so the Fourier transform $\hat{f}$ of f coincides away from a neighbourhood of 0 with a rapidly decreasing function, in fact with a function of the Schwartz class $S(\hat{\mathbb{R}})$. Now one can choose a sequence of compactly supported distributions $\hat{f}_n$ on $\hat{\mathbb{R}}$, coinciding near 0 with $\hat{f}$, such that $\hat{f}_n - \hat{f} \to 0$ in $S(\hat{\mathbb{R}})$. Then $f_n - f \to 0$ in $S(\mathbb{R})$, so $f_n \in S^0(\mathbb{R})$ and $f_n \to f$ in $S^0(\mathbb{R})$. In particular, $f_n \to f$ uniformly, so by the spectral theorem $f_n(D) \to f(D)$ in norm. Since $f_n(D) \in V$ by (2.2), $f(D) \in \overline{V}$.

(ii) If $\hat{f} \in C_c^\infty(\hat{\mathbb{R}})$, then $f \in \mathcal{S}(\mathbb{R})$. By (2.2), $f(D) \in \mathcal{V}$. But for any k, $D^k f(D)$ is bounded on L^2, so by standard elliptic estimates, $f(D)$ is a smoothing operator; hence $f(D) \in \mathcal{X}$.

(iii) It is easy to check that if $f \in C_0(\mathbb{R})$, then there is a sequence of functions f_n with $\hat{f}_n \in C_c^\infty(\hat{\mathbb{R}})$ such that $f_n \to f$ uniformly, so $f_n(D) \to f(D)$ is norm. Since $f_n(D) \in \mathcal{X}$, $f(D) \in \bar{\mathcal{X}}$. □

(2.4) <u>Remarks:</u> i) If $f \in S^0(\mathbb{R})$, then by (Taylor [15], Ch. XII), $f(D)$ is in fact a pseudo-differential operator of order 0. We will not need this fact, however.

ii) Notice that we can estimate the width of the support of the kernel of $f(D)$ in terms of the size of $\text{supp}(\hat{f})$.

Now, we shall construct the odd index. Recall that $\bar{\mathcal{X}}$ is an ideal in $\bar{\mathcal{V}}$, and consider the short exact sequence

(2.5) $0 \to \bar{\mathcal{X}} \to \bar{\mathcal{V}} \to \mathcal{Q} \to 0$

of C*-algebras. Choose any function $f \in S^0(\mathbb{R})$ such that $f(x) \to -1$ as $x \to -\infty$, $f(x) \to 1$ as $x \to \infty$. By (2.3), $f(D)$ is an element of $\bar{\mathcal{V}}$.

(2.6) <u>Lemma:</u> $\pi f(D)$ is an involution in $\mathcal{Q}$. It does not depend on the choice of f.

<u>Proof</u>: $f(D)^2 - 1 = g(D)$, where $g = f^2 - 1 \in C_0(\mathbb{R})$. Hence by (2.3), $g(D) \in \bar{\mathcal{X}}$, so $f(D)$ is an involution modulo $\bar{\mathcal{X}}$, so $\pi f(D)^2 = 1$. Similarly, if f_1 and f_2 are two choices for f, then $f_0 - f_1 \in C_0(\mathbb{R})$, so $f_0(D) - f_1(D) \in \bar{\mathcal{X}}$, and $\pi f_0(D) = \pi f_1(D)$. □

Since $\pi f(D)$ is an involution, it defines an element of $K_0(\mathcal{Q})$. Let ∂ denote the connecting homomorphism $K_0(\mathcal{Q}) \to K_1(\bar{X})$ in the long exact sequence of algebraic K-theory arising from (2.5).

(2.7) Definition: The odd index of D, odd-ind(D), is defined to be the element $\partial[\pi f(D)]$ of $K_1(\bar{X})$.

This odd index can be significant only on non-compact manifolds. This is a corollary of the following vanishing theorem for the odd index:

(2.8) Proposition: Suppose that the spectrum of D has a gap: i.e. that there is some real $x_0 \notin \sigma(D)$. Then odd-ind$(D) = 0$.

Proof: There is an interval $(x_0 - \varepsilon, x_0 + \varepsilon)$ in the resolvent set of D, and we can find a function $f \in S^0(\mathbb{R})$ such that $f(x) = -1$ for $x < x_0 - \varepsilon/_2$, $f(x) = +1$ for $x > x_0 + \varepsilon/_2$. But now $f(D)^2 = 1$ exactly, since $f = \pm 1$ on $\sigma(D)$. Therefore, $f(D)$ gives an element of $K_0(\bar{V})$, and odd-ind$(D) = \partial \circ \pi_*[f(D)] = 0$ by exactness of the K-theory sequence. □

§3. The operator on N

Recall that D is a first order self-adjoint differential operator on the bundle S. We will say that D is of Dirac type if S is a bundle of Clifford modules, and the principal symbol of D is given by Clifford multiplication. (Automatically then, D has propagation speed 1).

Suppose that D is an operator of Dirac type, and let $\underline{n}$ denote the unit normal covector field to N. Then $\sigma_D(\underline{n})$ is an anti-involution of the bundle $S_{|N}$, which therefore decomposes as a sum $S^+ \oplus S^-$ of $+i$ and $-i$ eigenbundles for $\sigma_D(\underline{n})$.

Let D_N denote a first order operator on $S_{|N}$ whose principal symbol is given by

$$\sigma_{D_N}(\underline{\xi}) = \sigma_D(i(\underline{\xi})),$$

where $i : T^*N \to T^*M$ is defined by the Riemannian metric. Since $i(\underline{\xi})$ is perpendicular to $\underline{n}$, $i(\underline{\xi})$ anticommutes with $\underline{n}$ in the Clifford algebra, and therefore D_N anticommutes (modulo operators of order zero) with the anti-involution $\sigma_D(\underline{n})$. Therefore, D_N restricts to an elliptic operator from S^+ to S^-.

(3.1) Definition: We call D_N "the" operator on N induced by the operator D on M.

(3.2) Example: Let D be the classical Dirac operator on M, so that the fibres of S are copies of the odd-dimensional spin representation, say of $Spin(2n+1)$. The fibres of $S_{|N}$ are representations of $Spin(2n)$; the inclusion $Spin(2n) \to Spin(2n+1)$ being induced by $TN \to TM$. Now under

Spin(2n) $\to$ Spin(2n+1) the spin representation Δ pulls back to $\Delta^+ \oplus \Delta^-$. These representations Δ^+ and Δ^- can be identified with the $\pm i$-eigenspaces of $\underline{n}$, considered as an element of the Clifford algebra. Thus we see that D_N is just the usual Dirac operator on N, whose index in the $\hat{A}$-genus $\hat{A}[N]$.

Our main result is the following:

(3.3) INDEX THEOREM: Let M be a complete oriented Riemannian manifold, $E = (M^+, M^-, N)$ a partition of M. Let D be an operator of Dirac type on a bundle S, and let odd-ind(D) $\in K_1(\bar{X})$ be defined as in §2. Let $\zeta_E^* : K_1(\bar{X}) \to \mathbb{C}$ be defined as in §1. Then

$$\zeta_E^*(\text{odd-ind } D) = \text{Index}(D_N) .$$

(The r.h.s. is the usual even index, which may be computed by the Atiyah-Singer formula).

We give a sketch of the proof, which will occupy §§4-8. In §4 we will prove the assertion of the theorem in the easiest case: $M = \mathbb{R}$, equipped with the usual metric. This reduces to the theory of Wiener-Hopf operators. In §5 we will extend this by means of a product construction to $N \times \mathbb{R}$ with product metric. In §6 we shall show that for a general manifold M, the index remains unchanged if the operator D is deformed to a product near N, and in §7 we shall prove a localisation theorem which says that ζ_E^* (odd-ind D) depends only on the geometry of a neighbourhood of N. We put these results together to obtain the general theorem in §8.

§4. The case $M = \mathbb{R}$

We will prove the theorem of §3 in the case $M = \mathbb{R}$, S = trival line bundle, $D = i\frac{d}{dx}$, $\mathcal{E} = (\mathbb{R}^+, \mathbb{R}^-, \{0\})$. We begin by calculating the odd index of D.

The group $\mathbb{R}$ acts on M by translation, and hence acts on the algebras $\mathcal{V}$, $\overline{\mathcal{V}}$, $\mathcal{X}$, and $\overline{\mathcal{X}}$. Thus we have a diagram

$$\begin{array}{ccccccccc} 0 & \to & \overline{\mathcal{X}} & \to & \overline{\mathcal{V}} & \to & \mathcal{Q} & \to & 0 \\ & & \uparrow & & \uparrow & & \uparrow & & \\ 0 & \to & \overline{\mathcal{X}}^{\mathbb{R}} & \to & \overline{\mathcal{V}}^{\mathbb{R}} & \to & \mathcal{Q}^{\mathbb{R}} & \to & 0 \end{array}$$

(the upper $\mathbb{R}$ denoting the set of fixed points). Since D is a translation-invariant operator, $f(D) \in \overline{\mathcal{V}}^{\mathbb{R}}$ for all $f \in S^0(\mathbb{R})$, and therefore $\text{odd-ind}(D) \in K_1(\overline{\mathcal{X}}^{\mathbb{R}})$.

(4.1) Proposition: a) The algebras $\mathcal{X}^{\mathbb{R}}$ and $\overline{\mathcal{X}}^{\mathbb{R}}$ are isomorphic to $C_c^\infty(\mathbb{R})$ and $C_0(\hat{\mathbb{R}})$ respectively, acting by convolution on $L^2(\mathbb{R})$, resp. multiplication on $L^2(\hat{\mathbb{R}})$.

b) The group $K_1(\overline{\mathcal{X}}^{\mathbb{R}})$ is isomorphic to $\mathbb{Z}$.

Proof: a) An element of $\mathcal{X}^{\mathbb{R}}$ is an operator given by a smooth convolution kernel $k(x,y)$ such that $k(x,y) = 0$ if $|x-y| > R$ and $k(x+t,y+t) = k(x,y)$; whence it follows that $k(x,y) = f(x-y)$, $f \in C_c^\infty(\mathbb{R})$. Taking the Fourier transform, we can identify $\mathcal{X}$ with a dense subalgebra of the Schwartz functions on $\hat{\mathbb{R}}$, acting by multiplication on $L^2(\hat{\mathbb{R}})$. By Plancherel's theorem, then, the completion of $\mathcal{X}$ is the C*-algebra $C_0(\hat{\mathbb{R}})$.

b) We have $K_1(\mathcal{X}^{\mathbb{R}}) \cong K_1(C_0(\hat{\mathbb{R}})) \cong K_c^1(\hat{\mathbb{R}}) \cong \mathbb{Z}$.

We can calculate an explicit generator: it is given by any $b \in C_0(\hat{\mathbb{R}})$ such that $|1 + \hat{b}| = 1$ and the map $1 + \hat{b} : \hat{\mathbb{R}} \to S^1$, which sends $\pm\infty$ to 1, has degree 1. We can take $b \in C_c^\infty(\mathbb{R})$ if we wish, so we see that $K_1(X^{\mathbb{R}}) \to K_1(\mathcal{X}^{\mathbb{R}})$ is surjective. □

(4.2) <u>Proposition</u>: The odd index of D is the generator of $K_1(\mathcal{X}^{\mathbb{R}})$.

<u>Proof</u>: Notice first that the spectrum of the operator D is canonically isomorphic to $\hat{\mathbb{R}}$. Thus, the isomorphism of $C_0(\hat{\mathbb{R}})$ with $\mathcal{X}^{\mathbb{R}}$ of proposition (4.1) is in fact given by $f \to f(D)$. Similarly, by the results of §2, there is an injection given by $f \to f(D)$ from $S^0(\hat{\mathbb{R}})$ to $\mathcal{V}^{\mathbb{R}}$. However, $\mathcal{V}^{\mathbb{R}}$ is closed in the C*-topology, which is equivalent to the uniform topology on functions f. Therefore, $\mathcal{V}^{\mathbb{R}}$ contains a subalgebra isomorphic (via the map $f \to f(D)$) to $C(\hat{\mathbb{R}}_{II})$, the continuous functions on the two-point compactification $\hat{\mathbb{R}}_{II}$ of $\hat{\mathbb{R}}$. Moreover, the function f used in §2 to define the odd index certainly belongs to $C(\hat{\mathbb{R}}_{II})$.

Now consider the diagram of exact sequences

$$\begin{array}{ccccccccc} 0 & \to & \mathcal{X}^{\mathbb{R}} & \to & \mathcal{V}^{\mathbb{R}} & \overset{\pi}{\to} & \mathcal{Q}^{\mathbb{R}} & \to & 0 \\ & & \uparrow & & \uparrow & & \uparrow & & \\ 0 & \to & C_0(\hat{\mathbb{R}}) & \to & C(\hat{\mathbb{R}}_{II}) & \to & \mathbb{C}\oplus\mathbb{C} & \to & 0 \end{array}$$

and the corresponding diagrams of K-theory maps. Since $f \in C(\hat{\mathbb{R}}_{II})$, we see that $[\pi f(D)] \in K_0(\mathbb{C} \oplus \mathbb{C})$.

Now the bottom exact sequence is an exact sequence of commutative C*-algebras, and its associated K-theory sequence is just the corresponding exact sequence in topological K-theory of the maximal ideal spaces:

$$\cdots \to K_c^0(\hat{\mathbb{R}}) \to K^0(\hat{\mathbb{R}}_{II}) \to K^0(2) \to K_c^1(\hat{\mathbb{R}}) \to K^1(\hat{\mathbb{R}}_{II}) \to \cdots .$$

Here 2 denotes a 2-point space. Now $\hat{R}_{II}$ is contractible, so $K^0(\hat{\mathbb{R}}_{II}) \cong \mathbb{Z}$, $K^1(\hat{\mathbb{R}}_{II}) \cong 0$. $K^0(2)$ is the direct sum of two copies of $\mathbb{Z}$, one of which is the image of $K^0(\hat{\mathbb{R}}_{II})$ and the other of which is generated by a vector bundle of dimension zero over $-\infty$, dimension one over $+\infty$. This is exactly $[\pi f(D)]$. But by exactness, $[\pi f(D)]$ maps to a generator of $K_c^1(\hat{\mathbb{R}}) \cong \mathbb{Z}$, as required.

There is a sign to check, to make sure that the generator of $K_c^1(\hat{\mathbb{R}})$ obtained is the one with winding number $+1$, not -1. To do this, notice that the winding number map $K_c^1(\hat{\mathbb{R}}) \to \mathbb{Z}$ can be described homologically as the pairing of the Chern character with the generator g of $H_1(\hat{\mathbb{R}})$ given by integration. The boundary $bg \in H_0(2)$ is then zero over $-\infty$, one over $+\infty$; so

$$\langle \mathrm{ch}(\text{odd-ind}(D)), g\rangle = \langle \mathrm{ch}[\pi f(D)], bg\rangle > 0 . \qquad \square$$

Now let E be the standard partition of $\mathbb{R}$. Then there is a diagram

$$\mathbb{Z} \cong K_1(\bar{X}^{\mathbb{R}}) \to K_1(\bar{X}) \xrightarrow{\zeta_E^*} \mathbb{C} .$$

(4.3) <u>Proposition</u>: ζ_E^* coincides with the winding number map on $K_1(\bar{X}^{\mathbb{R}})$.

<u>Proof</u>: It is enough to show that ζ^* maps the generator to 1. As we remarked above, the generator is given by some

$b \in C_c^\infty(\mathbb{R})$, acting by convolution, such that $|1 + \hat{b}| = 1$ and $1 + \hat{b}$ has winding number 1 as a map $\hat{\mathbb{R}} \to S^1$.

Then b defines an operator in $X^{\mathbb{R}}$, and therefore lies in the domain of the cyclic cocycle ζ. Since $1 + b$ is unitary, the formula for Connes' pairing (Connes [6], Ch.II., Prop. 15) is just $\zeta(b^*, b)$, where b^* is the adjoint of b considered as a convolution operator, i.e. $b^*(x) = \bar{b}(-x)$. (We have left out the normalizing constant of $\frac{1}{8\pi i}$ given by Connes). Now using the formula for ζ given in §1, we find that $\zeta(b^*, b)$ is the trace of an integral operator on $\mathbb{R}$ with kernel

$$k(x,z) = \begin{cases} 0 & \text{if } x \le 0 \text{ or } z \le 0 \\ & \text{otherwise} \\ \displaystyle\int_{y<0} (b(x-y)b^*(y-z) - b^*(x-y)b(y-z))\,dy & \end{cases}$$

Write then $I(u) = b(u)b^*(-u) - b^*(u)b(-u)$; the trace $\zeta(b^*, b)$ is then

$$\int_{x>0} dx \int_{y<0} dy\, I(x-y) .$$

Substituting $u = x - y$, we get

$$\int_0^\infty du \int_0^u dx\, I(u) .$$

This is

$$\int_0^\infty uI(u)du = \int_{-\infty}^\infty ub(u)b^*(-u)du$$

$$= \frac{1}{2\pi i}\int_{-\infty}^\infty \hat{b}'(\xi)\bar{\hat{b}}(\xi)d\xi \quad \text{by Plancherel's theorem}$$

$$= \text{winding number of } \hat{b} = 1, \quad \text{as asserted.} \quad \square$$

Combining these results, we see that (as asserted) the odd-index of D, which is the generator of $K_1(X^{\mathbb{R}})$, maps to 1 under the map ζ^*_E. Now let us consider the 'even operator' D_N; here N is 0-dimensional, and D_N is a linear map from a 1-dimensional vector space to a 0-dimensional space. Thus $\text{index}(D_N) = 1$, and the theorem of §3 is verified in this case. $\square$

§5. Proof for $\mathbb{R} \times N$

We will now consider the case that M is a Riemannian product manifold $\mathbb{R} \times N$, N being compact, and the partition E is induced from the canonical partition of $\mathbb{R}$ discussed in §4. Our operator D will be obtained from an operator on N as follows. Let D_N be a graded self-adjoint operator of Dirac type on N, that is, D_N operates on sections of a vector bundle $S_N = S_N^+ \oplus S_N^-$ over N, mapping S_N^+ to S_N^- and S_N^- to S_N^+. Pull back S_N to a bundle S over M in the obvious way, and define an operator D on S by

$$D = \begin{bmatrix} i\,\frac{d}{dx} & D_N \\ & \\ D_N & -i\,\frac{d}{dx} \end{bmatrix}$$

$$= \gamma_1 i\frac{d}{dx} + \gamma_2\, D_N\,,$$

where $\gamma_1 = \begin{bmatrix} 1 & 0 \\ 0 & -1 \end{bmatrix}$ and $\gamma_2 = \begin{bmatrix} 0 & 1 \\ 1 & 0 \end{bmatrix}$ are two of the Pauli matrices. D_N is then induced by D in the sense of §3. Notice in particular that if N is spin, this is exactly the relationship between the Dirac operators on M and on N.

We will prove the product formula for the odd index in this case by decomposing the space $L^2(S)$ according to the spectral theory of the operator D_N on the compact manifold N. We can think of $L^2(S)$ as the space of L^2 sections of the infinite-dimensional trivial vector bundle over $\mathbb{R}$ whose fibre is the space $L^2(S_N)$. Now we can decompose

$$L^2(S_N) \cong V_N \oplus W_N$$

where $V_N = \ker(D_N)$ is a finite dimensional subspace, and W_N is its orthogonal complement. Correspondingly we decompose

$$L^2(S) \cong V \oplus W$$

where V (resp. W) is the space of L^2 maps from $\mathbb{R}$ to V_N (resp. W_N). Let $p : L^2(S) \to V$ denote the orthogonal projection onto V.

(5.1) <u>Proposition</u>: a) p belongs to the algebra $\mathcal{U}(S)$.

b) The operator D commutes with p.

<u>Proof</u>: a) Clearly p is a bounded operator on L^2. Moreover, the operator p on $L^2(S) \cong L^2(\mathbb{R}) \otimes L^2(S_N)$ is equal to $1 \otimes q$, where $q : L^2(S_N) \to V_N$ is the orthogonal projection, which is a smoothing operator. Therefore p is given by a Schwartz kernel, the tensor product of the δ-function on the diagonal in $\mathbb{R} \times \mathbb{R}$ with the smoothing kernel of q on $N \times N$. Since N has finite diameter, the Schwartz kernel of p is supported within a finite distance of the diagonal.

b) This is clear, since $\frac{d}{dx} = \frac{d}{dx} \otimes 1$ commutes with $p = 1 \otimes q$, and $D_N p = pD_N = 0$ since D_N is self-adjoint. □

Let us now define $\mathcal{U}^p$ to be the commutant of p in $\mathcal{U}$, $\mathcal{X}^p = \mathcal{X} \cap \mathcal{U}^p$. By definition, p is a central projection in $\mathcal{U}^p$, so $\mathcal{U}^p \cong p\mathcal{U}^p \oplus (1-p)\mathcal{U}^p$, $\mathcal{X}^p \cong p\mathcal{X}^p \oplus (1-p)\mathcal{X}^p$. Moreover, if $f \in S^0(\mathbb{R})$, then by (5.1)(b) $f(D) \in \mathcal{U}^p$, i.e. $f(D)$ commutes with p. Therefore, if we consider the exact sequences

$$\begin{array}{ccccccccc} 0 & \to & \mathcal{X} & \to & \mathcal{V} & \xrightarrow{\pi} & \mathcal{Q} & \to & 0 \\ & & \uparrow & & \uparrow & & \uparrow & & \\ 0 & \to & \mathcal{X}^p & \to & \mathcal{V}^p & \xrightarrow{\pi} & \mathcal{Q}^p & \to & 0 \end{array}$$

We see that odd-ind(D) can be considered as an element of $K_1(\mathcal{X}^p) \cong K_1(p\mathcal{X}^p) \oplus K_1((1-p)\mathcal{X}^p)$.

Let us choose the function $f \in S^0(\mathbb{R})$ used in defining the odd index as follows: $f \to -1$ at $-\infty$, $f \to 1$ at ∞, $f(0) = 0$, and $f(\lambda) = \pm 1$ for all $|\lambda| > |\lambda_0|$, λ_0 the non-zero eigenvalue of D_N of smallest absolute value.

The odd index of D is then defined as in §2 from $[\pi f(D)]$ in $K_0(\mathcal{Q}^p)$.

(5.2) Proposition: The component of odd-ind(D) in $K_1((1-p)\mathcal{X}^p)$ is zero.

Proof: We argue as in the vanishing theorem of §2. It is enough to show that $(1-p)f(D)$ is an involution in $(1-p)\mathcal{V}^p$, so giving a class in K_0 of this algebra.

Now

$$D = \gamma_1 i \frac{d}{dx} + \gamma_2 D_N .$$

Therefore

$$D^2 = -\gamma_1^2 \left(\frac{d}{dx}\right)^2 + \gamma_2^2 D_N^2 + i(\gamma_1\gamma_2 + \gamma_2\gamma_1) \frac{d}{dx} D_N$$

$$= -\frac{d^2}{dx^2} + D_N^2 .$$

Hence, for any $u \in \mathrm{dom}(D^2) \cap \mathrm{range}(1-p)$

$$\langle D^2u,u\rangle = \langle -\frac{d^2}{dx^2}u,u\rangle + \langle D_N^2 u,u\rangle$$

$\geq \lambda_0^2 \|u\|^2$, where λ_0 is the least (in abs. value) nonzero eigenvalue of D_N.

Therefore the spectrum of D acting on the Hilbert space W does not meet the interval $(-\lambda_0,\lambda_0)$. Since $f = \pm 1$ this interval, $f(D)$ is an involution on W, i.e. $(1-p)f(D)$ is an involution in $(1-p)\overline{\mathcal{U}}^p$, as asserted. □

Now consider the odd index in $K_1(p\overline{X}^p)$. Here $p\overline{X}^p(M)$ is just isomorphic to the algebra $\overline{X}(\mathbb{R},S_1)$, where the coefficient bundle S_1 is the trivial bundle with fibre $\ker(D_N)$, and D acts by the matrix

$$\begin{bmatrix} i\frac{d}{dx} & 0 \\ 0 & -i\frac{d}{dx}\end{bmatrix};$$

acting as $i\frac{d}{dx}$ on $\ker(D_N^+)$ and $-i\frac{d}{dx}$ on $\ker(D_N^-)$. By the results of §4, now, on $\mathbb{R}$, $\zeta_E^*(\text{odd-ind } i\frac{d}{dx}) = 1$. Thus we get

$$\zeta_E^*(\text{odd-ind } D)$$

$$= \zeta_E^*(p_*(\text{odd-ind } D)) \qquad \text{(by (5.2))}$$

$$= (\dim\ker D_N^+)\zeta_E^*(\text{odd-ind}(i\frac{d}{dx})) + (\dim\ker D_N^-)\zeta_E^*(\text{odd-ind}(-i\frac{d}{dx}))$$

$$= \dim\ker D_N^+ - \dim\ker D_N^-$$

$$= \text{Index } D_N .$$

Thus we obtain

(5.3) Proposition: The Index Theorem of §3 holds if M is a product manifold.

§6. Deformation

In this section we shall prove that the odd index is a homotopy invariant in a suitable sense. This will then be used to reduce the index theory of a general operator to the case studied in §5.

Before studying deformations of the operators, we must briefly consider the effect of deformations of the Riemannian metric on the algebras $\mathcal{V}$, $\mathcal{X}$, $\overline{\mathcal{V}}$, and $\overline{\mathcal{X}}$. The Riemannian metric enters into their definition in two places: in measuring the width of the support of the Schwartz kernel, and in defining the C^*-norm in which $\mathcal{V}$ and $\mathcal{X}$ are completed to obtain $\overline{\mathcal{V}}$ and $\overline{\mathcal{X}}$.

Suppose now that the Riemannian metric is deformed within the quasi-isometry class. (In particular, a compactly supported deformation will have this property). Then distances are deformed by at most some constant factor. Moreover, the topology of the vector space $L^2(S)$ remains unchanged, though its norm is replaced by an equivalent one. Therefore the algebras $\mathcal{V}$ and $\mathcal{X}$ remain unchanged. As locally convex topological algebras, $\overline{\mathcal{V}}$ and $\overline{\mathcal{X}}$ also remain unchanged, though their norms and $*$-operations are replaced by equivalent ones. Therefore, $K_1(\overline{\mathcal{X}})$ remains unchanged under deformation of the metric.

Now let us consider the following situation: D_t is a continuous family of first order self-adjoint operators on the manifold M, with $D_t \equiv D_0$ outside a compact subset of M.

(6.1) Proposition: In the above situation, odd-ind$(D_t) \in K_1(\bar{X})$ is independent of t.

Proof: Let $f(x) = \frac{2x}{\sqrt{3+x^2}}$; then $f \in S^0(\mathbb{R})$, $f(x) \to \pm 1$ as $x \to \pm\infty$. Therefore, odd-ind(D_t) can be defined as in §2 as $\partial[\pi f(D_t)]$, where $\pi : \bar{V} \to \bar{V}/\bar{X} \cong \mathcal{Q}$ and $\partial : K_0(\mathcal{Q}) \to K_1(\bar{X})$ We claim that $f(D_t)$ is norm continuous in t. Then, by the homotopy invariance of the K-theory of Banach algebras (Blackadar [3], 4.3.3), we will find that $[\pi f(D_t)]$ represents an element independent of t in $K_0(\mathcal{Q})$.

To prove the continuity, let us argue as follows. The operator $D_0 - D_t$ is first order and compactly supported, and tends to zero as $t \to 0$. Therefore, the usual elliptic estimates tell us that given any $\delta > 0$ there exists t_0 s.t. $\forall\, t \in [-t_0, t_0]$,

$$\|D_0 s - D_t s\| \leq \delta(\|s\| + \|D_0 s\|) . \tag{6.2}$$

Now let A denote the operator $2D_t(3+D_0^2)^{-\frac{1}{2}}$. The estimates (6.2) show that A is a well-defined, bounded operator. In fact

$$\|A - f(D_0)\| = \|2(D_t - D_0)(3 + D_0^2)^{-\frac{1}{2}}\| \leq 3\delta .$$

Now consider

$$A - f(D_t) = f(D_t)[(3 + D_t^2)^{\frac{1}{2}}(3 + D_0^2)^{-\frac{1}{2}} - 1] .$$

Since $\|f(D_t)\|$ is certainly bounded, it is enough now to prove that $(3 + D_t^2)^{\frac{1}{2}}(3 + D_0^2)^{-\frac{1}{2}} - 1$ is of small norm. Now

$$(3+D_t^2)(3+D_0^2)^{-1} - 1 = (D_t^2-D_0^2)(3+D_0^2)^{-1} .$$

Now again from standard elliptic estimates there exists b_1 s.t. $\forall\, t \in [-t_1, t_1]$

$$\|D_0^2 s - D_t^2 s\| \le \delta(\|s\| + \|D_0^2 s\|) .$$

Therefore $\|(D_t^2 - D_0^2)(3 + D_0^2)^{-1}\| < \delta$ for sufficiently small t. Let $B = (D_t^2 - D_0^2)(3 + D_0^2)^{-1}$, so that

$$A - f(D_t) = f(D_t)((1+B)^{\frac{1}{2}} - 1) .$$

If $\delta < 1$ then we have the binomial series (convergent in the Banach algebra $\overline{X}$)

$$A - f(D_t) = f(D_t)(\tfrac{1}{2}B - \tfrac{1}{8}B^2 + \tfrac{1}{16}B^3 - \dots)$$

so

$$\|A - f(D_t)\| \le \|f(D_t)\|((1 + \delta)^{\frac{1}{2}} - 1)$$

$$\le (\text{const.})\delta .$$

Since also $\|A - f(D_0)\| \le 2\delta$, we get

$$\|f(D_0) - f(D_t)\| \le (\text{const.})\delta,$$

as required. □

7. Localization

Our aim in this section is to prove that $\zeta^*_{\mathcal{E}}$(odd-ind D) depends only on the geometry of M and D in a neighbourhood of the partitioning manifold N. We do this by a version of the heat equation method, getting local asymptotics for $\zeta^*_{\mathcal{E}}$(odd-ind D). Since odd-ind(D) depends on the boundary map in the algebraic K-theory exact sequence, we must begin by considering this map. Let us consider abstractly a short exact sequence of Banach algebras (with A, C unital)

$$0 \to J \to A \to C \to 0 .$$

The odd index uses $\partial : K_0(C) \to K_1(J)$. This map is constructed explicitly by using Bott periodicity (cf. Blackadar [3] Ch.9) as follows:

Step 1: Given a projection e in C (more generally, e could be a projection in a matrix algebra over C; but we will not need this) that defines an element of $K_0(C)$, we form a loop of invertibles α in C, defined by

$$\alpha(\theta) = (1 - e) + e.\exp(2\pi i\theta) = \exp(2\pi i\theta .e) .$$

Step 2: The loop $\begin{bmatrix} \alpha & 0 \\ 0 & \alpha^{-1} \end{bmatrix}$ of invertibles in $M_2(C)$ can be lifted to a loop β of invertibles in $M_2(A)$.

Step 3: The conjugate $\beta \begin{bmatrix} 1 & 0 \\ 0 & 0 \end{bmatrix} \beta^{-1}$ gives a loop γ of projections in $M_2(J^+)$. (Here J^+ denotes J with an adjoined unit).

So far we have not mentioned the Banach algebra structure at all; no analysis has been involved. Analysis enters in the next step:

Step 4: we find an invertible element of $M_2(J^+)$ which gives the 'holonomy' of the loop γ (cf. Blackadar [3], theorem 8.2.2) and this holonomy defines the desired element of $K_1(J)$. This uses the lemma that nearby projections in a Banach algebra are conjugate.

One way to evaluate this holonomy is the following. Let $e_\theta, \theta \in \mathbb{T}$, be a diferentiable loop of projections in a Banach algebra B. Define a loop u_θ of invertible elements by solving the differential equation

$$u_0 = 1, \quad \dot{u}_\theta = -[e_\theta, \dot{e}_\theta]u_\theta .$$

If we define v_θ to be the solution of the differential equation

$$v_0 = 1, \quad \dot{v}_\theta = v_\theta[e_\theta, \dot{e}_\theta],$$

then it is easy to verify that v_θ is the inverse of u_θ, and that

$$\frac{d}{d\theta}(v_\theta e_\theta u_\theta) = 0 .$$

Hence

$$e_\theta = u_\theta e_0 u_\theta^{-1}$$

and the holonomy is given by $u_{2\pi}e_{2\pi} + 1 - e_{2\pi}$. (Notice that $e_{2\pi} = e_0$ commutes with $u_{2\pi}$).

Now we are interested in pairing the odd index with a cyclic 1-cocycle. It turns out that we can pair a cyclic 1-cocycle directly with Step 3 data; that is, with a differentiable loop of projections:

(7.1) Lemma : Let B be a Banach algebra and let $e_\theta, \theta \in \mathbb{T}$, be a differentiable loop of projections in B. Let ζ be a continuous cyclic 1-cocycle on B. Then the result of evaluating ζ on the K_1 class $[e_\theta]$ is given by the integral formula

$$-\int_0^{2\pi} \zeta([e_\theta, \dot{e}_\theta], e_\theta)\, d\theta \quad ,$$

where the dot denotes differentiation with respect to θ.

Proof: According to the formula of Connes [6], the value of the pairing is given by $\zeta(u,u^{-1})$, where u is the holonomy; in the notation above, $U = u_{2\pi}e_{2\pi}$. Let us write $E = e_{2\pi}$. Now consider

$$\frac{d}{d\theta}\{\zeta(u\,E, Eu^{-1})\} = \zeta(uE, Eu^{-1}[e,\dot{e}]) - \zeta([e,\dot{e}]uE, Eu^{-1})$$

(we omit the subscripts θ)

$$= -\zeta([e,\dot{e}], uE.Eu^{-1})$$

(cyclic property)

$$= -\zeta([e,\dot{e}], e) \ .$$

The result follows. □

(7.2) Lemma: Let $\mathcal{B}$ be any algebra over $\mathbb{C}$, and let $[e_\theta]$ be a loop of projections in $\mathcal{B}$, whose values lie in a finite-dimensional subspace of $\mathcal{B}$, and which is differentiable there. Let ζ be a cyclic 1-cocycle on $\mathcal{B}$.

The the formula of Lemma (7.1) defines a number which is equal to the result of applying ζ_* to $[e_\theta]$ in $K_1(B)$ for any Banach algebra B containing $\mathcal{B}$ to which ζ extends continuously.

Proof: Immediate from (7.1).

The loop of projections obtained from the odd index construction is of this sort. Specifically, suppose now that J is an ideal in an unital algebra A over $\mathbb{C}$, and that $P \in A$ is an idempotent modulo J. Then let us define loops $\tilde{P}$ and $\tilde{Q}$ in A by

$$\left.\begin{aligned} \tilde{P}(\theta) &= (1-P) + P\exp(2\pi i\theta) \\ \tilde{Q}(\theta) &= (1-P) + P\exp(-2\pi i\theta) \,. \end{aligned}\right\} \tag{7.3}$$

Then a loop β in $H_2(A)$ satisfying the requirements of Step 2 above can be defined by

$$\beta(\theta) = \begin{bmatrix} (2-\tilde{P}\tilde{Q})\tilde{P} & \tilde{P}\tilde{Q}-1 \\ 1-\tilde{Q}\tilde{P} & \tilde{Q} \end{bmatrix}$$

(cf. Milnor [12], §4). Step 3 then leads to a loop of projections in J^+,

$$\gamma(\theta) = \begin{bmatrix} 1-(1-\tilde{P}\tilde{Q})^2 & \tilde{Q}(\tilde{P}\tilde{Q}-1) \\ -(\tilde{P}\tilde{Q}-1)(\tilde{P}\tilde{Q}-2)\tilde{P} & (1-\tilde{Q}\tilde{P})^2 \end{bmatrix} \tag{7.4}$$

This loop $\gamma(\theta)$ clearly satisfies the conditions of Lemma (7.2). Therefore, $\zeta_*(\text{odd-ind } P)$ can be computed from the loop $\gamma(\theta)$ by the formula

$$\zeta_*(\text{odd-ind } P) = \int_0^{2\pi} \zeta([\gamma(\theta),\dot{\gamma}(\theta)],\gamma(\theta))d\theta \,.$$

In our geometrical situation, we are considering the ideal X in the algebra U of uniform operators. The approximate projection P is given by $\frac{1}{2}(f(D) + 1)$, where f is a function of the sort considered in (2.6). A priori, $P \in \bar{\mathcal{V}}$; but by (2.2), if the Fourier transform of f is compactly supported, $P \in \mathcal{V}$. The deviation $P^2 - P$ of P from idempotence is $\frac{1}{4}(f(D)^2 - 1)$; by suitable choice of f we can arrange that this belongs to $\mathcal{X}$:

(7.5) <u>Lemma</u>: Given any $R > 0$, there exists $f \in S^0(\mathbb{R})$, with $f(x) \to \pm 1$ as $x \to \pm\infty$, such that $f^2 - 1 = g \in \mathcal{S}(\mathbb{R})$, $\mathrm{Supp}(\hat{f}) \subset [-R,R]$, and $\mathrm{supp}(\hat{g}) \subset [-2R,2R]$.

<u>Proof</u>: Choose a smooth, non-negative bump function $h \in (\mathbb{R})$ with $\mathrm{supp}(h) \subset [-R,R]$, $\int h = 2$, and define f by

$$f(x) = -1 + \int_{-\infty}^{x} h(y)dy .$$

Then $i\xi\hat{f}(\xi) = \hat{h}(\xi)$, so that $\mathrm{supp}(f) \subset [-R,R]$ also. Clearly $\hat{g} = \hat{f}*\hat{f} - \delta$ has support in $[-2R,2R]$ as a distribution. However, it is easy to check directly that $g \in S(\mathbb{R})$, so that g is in fact a smooth function. □

By (2.3), if we choose f as in this lemma, then $p^2 - p \in \mathcal{X}$, so we are in the situation discussed above.

(7.6) <u>Lemma</u>: Given any $R > 0$, the odd index of D can be represented by a loop $\gamma(\theta)$ of projections in $\mathcal{X}^+$ whose Schwartz kernels are supported within distance R of the diagonal in $M \times M$. Moreover, the value of such a kernel at $(x,y) \in M \times M$ depends only on the geometry of (M,S) in a 2R-neighbourhood of x.

<u>Proof</u>: Immediate from (7.4), (7.5) and the remark after (2.3).

(7.7) Theorem: (Localization) Let $E = (M^+, M^-, N)$ be a partition of M. Then for any $R > 0$, $\zeta_E^*(\text{Odd-ind } D)$ depends only on the geometry of (M,S) in an R-neighbourhood of N.

Proof: Recall that

$$\zeta_E(A,B) = \mathrm{Tr}(-\pi A(1-\pi).(1-\pi)B\pi + \pi B(1-\pi).(1-\pi)A\pi) \quad .$$

If the kernel of A is supported within distance R of the diagonal, then the kernels of $\pi A(1-\pi)$ and $(-\pi)A\pi$ are supported within an R-neighbourhood of $N \times N$ in $M \times M$; in fact they are the restrictions to $M^+ \times M^-$ (resp. $M^- \times M^+$) of the Schwartz kernel of A. Thus for any $R > 0$, one can represent odd-ind(D) by a loop $\gamma(\theta)$ of projections which has the property that for each θ

$$\zeta_\varepsilon([\gamma(\theta),\dot{\gamma}(\theta)],\gamma(\theta))$$

depends only on the geometry of (M,S) in a R-neighbourhood of N. But now the formula (7.1)

$$\zeta_\varepsilon^*(\text{odd-ind } D) = -\int_0^{2\pi} \zeta_\varepsilon([\gamma(\theta),\dot{\gamma}(\theta)],\gamma(\theta))d\theta$$

gives the desired result. □

§8. Proof of the theorem

We will now use the results of §§4-7 to complete the proof of our main theorem, Theorem (3.3). Thus, let M be a complete oriented Riemannian manifold, equipped with a partition $E = (M^+, M^-, N)$. Let D be an operator of Dirac type on a bundle S. Then we most prove that

$$\zeta_E^*(\text{odd-ind } D) = \text{Index } D_N . \tag{8.1}$$

Let M' be the manifold $N \times \mathbb{R}$, equipped with the obvious partition E'. We equip M' with the bundle S', which is the pull-back of S under $M' \to N \to M$.

By the Tubular Neighbourhood Theorem, there is a neighbourhood u of N in M that is identified with $N \times (-\varepsilon, \varepsilon)$, for some $\varepsilon > 0$. Then U is diffeomorphic to $N \times (-\varepsilon, \varepsilon) \subset M'$, and this diffeomorphism maps the restriction to U of S to the corresponding restriction of S'. Therefore, the operator D on U can be transported to a differential operator D_1 on $N \times (-\varepsilon, \varepsilon)$.

Let D_2 be the operator on $N \times (-\varepsilon, \varepsilon)$ given by

$$D_2 = \gamma_1 \, i \frac{d}{dx} + \gamma_2 D_N \tag{8.2}$$

as in §5. The principal symbols of D_1 and D_2 then coincide on N. Let ϕ be a bump function on M', equal to 1 on a neighbourhood of N in M' and supported within $N \times (-\varepsilon, \varepsilon)$. Let g_1 be the metric on $N \times (-\varepsilon, \varepsilon)$ transported from M, and let g_2 be the product metric on $N \times (-\varepsilon, \varepsilon)$. Finally, let

$$g' = \phi g_1 + (1 - \phi) g_2 ,$$

define a metric on S' in the analgous way, and let

$$D' = X + X^*$$

where the formal adjoint X^* is worked out relative to the metric g, and $X = \phi^{\frac{1}{2}}D_1\phi^{\frac{1}{2}} + (1-\phi)^{\frac{1}{2}}D_2(1-\phi)^{\frac{1}{2}}$. Then the following things are true:

(8.3) g' is a complete metric on M'.

(8.4) D' is a self-adjoint first-order operator on S' with bounded propagation speed.

(8.5) There is a neighbourhood of N in M' that can be naturally identified with a neighbourhood of N in M by an identification preserving the metrics and taking S' to S, D' to D.

(8.6) D' is homotopic to D_2, via a homotopy of compact support that also deforms g' to g_2.

For tho homotopy (8.6), just replace ϕ by $(1-t)\phi$, $0 \leq t \leq 1$, in the formulae above.

Now by (8.5) and (7.7),

$$\zeta^*_E(\text{odd-ind } D) = \zeta^*_{E'}(\text{odd-ind } D') .$$

By (8.6) and (6.1),

$$\zeta^*_{E'}(\text{odd-ind } D') = \zeta^*_{E'}(\text{odd-ind } D_2) .$$

By (5.3)

$$\zeta^*_{E'}(\text{odd-ind } D_2) = \text{index } (D_N) .$$

The result follows. □

§9. A consequence

Among the many results on scalar curvature proved by Gromov and Lawson [9] one finds the following:

Theorem: ([9], (6.13) (iv)). Suppose that Y is a compact spin manifold with $\hat{A}(Y) \neq 0$. Then $Y \times \mathbb{R}$ carries no complete metric of postive scalar curvature.

From our index theorem, we get the following result:

(9.1) Theorem: Let X be a noncompact spin manifold, and suppose that X is disconnected by a compact hypersurface Y with $\hat{A}(Y) \neq 0$. Then X cannot carry a complete metric of uniformly positive scalar curvature.

Proof: Suppose X to possess such a metric. Then the Lichnerowicz-Weitzenbock formula for the Dirac operator D gives

$$\langle D^2 s, s\rangle \geq \tfrac{1}{4}K > 0.$$

Hence

$$\sigma(D) \cap (-\tfrac{1}{2}K^{\frac{1}{2}}, \tfrac{1}{2}K^{\frac{1}{2}}) = 0 .$$

Thus $\sigma(D)$ has a gap near 0. By (2.8), odd-ind$(D) = 0$, so $\zeta_E^*(\text{odd-ind } D) = 0$, where E is the partition given by Y. But by our main theorem,

$$\zeta_E^*(\text{odd-ind } D) = \text{ind } D_Y = \hat{A}(Y) . \qquad \square$$

References

1 M.F.Atiyah Bordism and cobordism, *Math. Proc. Camb. Phil. Soc.* **57**(1961), 200-208.

2 P.Baum and R.G.Douglas K-homology and index theory, *AMS Proc. Symp. Pure Math.* **38**(1982), 117-174.

3 B.Blackadar K-theory for operator algebras, Springer, 1987.

4 P.R.Chernoff Essential selfadjointness of powers of generators of hyperbolic equations, *J. Fctl. Anal.* **12**(1973), 401-414.

5 A.Connes Sur la theorie noncommutative de l'integration, *Springer Lecture Notes* **725**, 19-143.

6 A.Connes Non-commutative differential geometry, Chapter I, *Publ. Math. de l'IHES* **62**(1985), 41-144.

7 A.Connes The transverse fundamental class of a foliation, IHES preprint (1984).

8 B.L.Feigin and B.L.Tsygan On the cohomology of Lie algebras of generalized Jacobi matrices, *Funktsional Anal. i Prilozhen* **17**(1983), 86-87.

9 M.Gromov and B.Lawson Positive scalar curvature and the Dirac operator, *Publ. Math. de l'IHES* **58**(1983), 83-196.

10 M.Karoubi K-Theory. Springer, 1978.

11 G.G.Kasparov K-theory, group C*-algebras, and higher signatures. Preprint, Chernogolovka, 1981.

12 J.Milnor Algebraic K-theory, Princeton, 1971.

13 J.Roe An index theorem on open manifolds I, *J. Diff. Geom.*, to appear.

14 L.Schwartz Théorie des distributions a valeurs vectorielles. *Ann. de l'Institut Fourier* **7**(1957), 1-142.

15 M.Taylor Pseudo-differential operators, Princeton, 1982.

Cyclic Cohomology of Algebras of Smooth Functions on Orbifolds

by
Antony Wassermann
(University of Liverpool and University of California, Berkeley)

In this informal report† I will present several principles that can be used to compute the cyclic cohomology of various smooth algebras. These include

(a) $C^\infty(X)$, with X a smooth manifold (but by a different method to that of Alain Connes)
(b) $C^\infty(X) \rtimes G$, where G is a finite group acting by diffeomorphisms on X
(c) $C^\infty(X)^G$ or more generally smooth functions on an orbifold
(d) $C^\infty(X)$, where X is a smooth manifold with boundary or even corners
(e) $\mathcal{S}(G)$, the convolution algebra of Schwartz functions on a reductive Lie group, including in particular the case $G = \mathbf{R}^n$.

This last example was in fact the main motivation for this work since at the time it was done (in 1984) I was working on a conjecture of Connes concerning a generalisation of the Connes–Moscovici Index Theorem (see [**7**]). Here one was interested in computing the pairing between certain specific elements of the cyclic cohomology and the K–theory of $\mathcal{S}(G)$. The cyclic cocycles were defined by group cocycles, or equivalently by invariant forms on the homogeneous space G/K via the van Est isomorphism; while the elements of K–theory were represented by abstract indices of twisted Dirac operators, exactly as in [**20**]. The rough scheme of the computation was to find a concrete 'heat kernel' formula for a projection representing the abstract index, substitute it into the cocycle formula and then do a Getzler rescaling to obtain the answer in the limit as Planck's constant approached zero. To guarantee the existence of such a limit, it was necessary to develop a pseudodifferential calculus for invariant pseudodifferential operators on G/K with spin coefficients completely analogous to that introduced by Ezra Getzler in [**9**] for supermanifolds. I am still in the process of writing up these calculations.

While musing over these problems, it occurred to me that it would be interesting to compute directly what the cyclic cohomology of the Schwartz algebra $\mathcal{S}(G)$ actually was, since one already knew the answer for its K–theory. The basic starting point was again the Plancherel decomposition of Jim Arthur and Harish-Chandra, but this time in its full

† This paper is a slightly expanded version of a talk given at the L.M.S. Symposium held at the University of Durham in July, 1987.

force, that is for the Schwartz algebra rather than the larger reduced C* algebra. Roughly speaking this decomposition shows that the Schwartz algebra is topologically isomorphic to a countable direct sum (with certain decay conditions on the components) of algebras of the form

$$(\mathcal{S}(\mathbb{R}^m)\hat{\otimes}\mathcal{K}_\infty)^\Gamma.$$

Here the group Γ is a finite reflection group acting on the euclidean space $\mathbb{R}^m$, $\mathcal{K}_\infty$ is isomorphic to the algebra of smoothing operators consisting of infinite matrices with coefficients of rapid decay, and the action of Γ on the tensor product is cocycle conjugate to the obvious tensor product action, trivial on the second factor. With such a description of $\mathcal{S}(G)$ at hand, the computation of its cyclic cohomology can naturally be divided into a series of simpler steps. Thus one first treats separately the problems of computing the cyclic cohomology of crossed products, fixed point algebras, $\mathcal{S}(\mathbb{R}^m)$, etc., together with more general questions of Morita invariance (for infinite dimensional matrices over a given algebra), additivity (for countable sums with suitable decay conditions), actions of derivations, and localisation properties of Hochschild cohomology. Rather than give a detailed discussion of how all these problems are tackled, I shall instead give a list of seven fairly general principles, illustrating how they may be applied in diverse situations. Full details will appear in **[21]** where a more complete list of references can be found. (We apologize in advance for any ommissions from a list which we unavoidably have had to keep short.) It goes almost without saying that our principal reference is Chapter II of Connes' book **[6]**, from which we shall adopt the notation and terminology without further comment.

I. Use of Multiplier Derivations: Renormalisation

Let $\mathcal{A}$ and $\mathcal{B}$ be algebras, $a \mapsto \overline{a}$ a homomorphism of $\mathcal{A}$ into $\mathcal{B}$ and $\delta : \mathcal{A} \to \mathcal{B}$ a derivation of $\mathcal{A}$ into $\mathcal{B}$ with respect to this homomorphism, so that $\delta(a_1a_2) = \delta(a_1)\overline{a_2} + \overline{a_1}\delta(a_2)$ for all $a_1, a_2 \in \mathcal{A}$. (If $\mathcal{A}$ and $\mathcal{B}$ are in addition locally convex topological algebras, we shall of course require that ϕ and δ be continuous.) Since δ is, at least morally, an infinitesimal homomorphism of $\mathcal{A}$ into $\mathcal{B}$, it naturally induces maps δ^* that carry contravariant objects associated with $\mathcal{B}$ into contravariant objects associated with $\mathcal{A}$. In particular on $H^n(\mathcal{B},\mathcal{B}^*)$ or $H^n_\lambda(\mathcal{B})$ we have the *Lie derivative*

$$\delta^*\phi(a^0,a^1,\ldots,a^n) = \sum_{i=0}^{n}\phi(\overline{a^0},\ldots,\overline{a^{i-1}},\delta a^i,\overline{a^{i+1}},\ldots,\overline{a^n}),$$

which from a Hochschild or cyclic cocycle ϕ on $\mathcal{B}$ produces the same type of cocycle $\delta^*\phi$ on $\mathcal{A}$. Of particular interest is the case where $\mathcal{A}$ and $\mathcal{B}$ coincide, the homomorphism is the identity automorphism of $\mathcal{A}$ and δ is just a derivation of $\mathcal{A}$ into itself. δ is then an infinitesimal automorphism of $\mathcal{A}$ and, for the special case $\mathcal{A} = C^\infty(X)$, δ will just be a vector field on X, with δ^* giving the usual Lie derivative on currents. For simplicity we will stick to the simpler case of (ordinary) derivations of $\mathcal{A}$, although everything we shall do will apply in the more general setting of derivations between two algebras after applying rather obvious modifications.

Now if $\delta_a(x) = ax - xa$ is an inner derivation of $\mathcal{A}$, Connes has given a one-line argument in the proof of Proposition 5 in [6] to show that δ_a^* acts trivially on $H_\lambda^n(\mathcal{A})$. Indeed $\delta_a^*\phi = b(A\phi_a)$ where

$$\phi_a(a^1,\ldots,a^n) = \phi(a,a^1,\ldots,a^n). \qquad (\dagger)$$

There is a similar result for Hochschild cocycles, with a formula which agrees with the above formula on cyclic cocyles: $\delta_a^*\phi = b\psi$ where

$$\psi(a^0,\ldots,a^{n-1}) = \sum_{j=1}^{n}(-1)^j\phi(a^0,\ldots,a^{j-1},a,a^j,\ldots,a^{n-1}). \qquad (\ddagger)$$

Among other things, these results can be used to prove the invariance of Hochschild and cyclic cohomology of *unital* algebras under inner automorphisms. One then deduces the Morita invariance and additivity of these cohomology theories, again with the restriction that the algebras be unital. These properties may be established essentially by the same type of argument that one would use to prove the analogous results in algebraic K–theory. To prove Morita invariance and additivity, it suffices to know that any inner automorphism of a unital algebra implemented by an element of square one acts trivially on Hochschild or cyclic cohomology. This follows exactly as in the proof of Proposition 5 of [6] by noting that any such element may trivially be written as an exponential, so that one simply has to integrate equations (†) and (‡). In fact if u has square one, then we have $u = -i\exp(i\pi u/2)$ and so we may write

$$Ad(u)^*\phi - \phi = \int_0^{\pi/2}\frac{d}{ds}(\alpha_s^*\phi)ds = \int_0^{\pi/2} i\alpha_s^*\delta_u^*\phi\, ds = bA(\int_0^{\pi/2} i\alpha_s^*\phi_u ds),$$

where $\alpha_t = Ad(\exp(iut))$. (The resulting expression on the right hand side in fact turns out to be rational.) To prove Morita invariance, that is to show that the natural map of $H_\lambda^*(\mathcal{A})$ into $H_\lambda^*(M_n(\mathcal{A}))$ is an isomorphism, one just places $M_n(\mathcal{A}) = \mathcal{A}\otimes M_n(\mathbb{C})$ as a corner in $\mathcal{B} = \mathcal{A}\otimes M_n(\mathbb{C})\otimes M_n(\mathbb{C})$ and considers the inner automorphism of $\mathcal{B}$ obtained by flipping its last two factors. To prove additivity, one also uses an auxiliary algebra: if there are n summands $\mathcal{A}_i$, one introduces the algebra $\mathcal{B} = M_n(\mathcal{A}_1\oplus\cdots\oplus\mathcal{A}_n)$ and places $\oplus\mathcal{A}_i$ in it as a corner. One then uses the obvious inner automorphism of $\mathcal{B}$ of period two that spreads this corner out over the diagonal, so that the summand $\mathcal{A}_i$ ends up in the ith position on the diagonal.

Now one wishes to obtain similar results when the algebras are no longer assumed to be unital; in particular one would like analogous results on the action of a derivation implemented by an element a which is just a *multiplier* of $\mathcal{A}$. We need to impose some extra conditions on the topological algebra in order to get anywhere. As we shall see, these conditions are quite natural. We require that $\mathcal{A}$ has:

(i) a two–sided approximate identity
(ii) a *Laplacian*, Δ, that is an invertible multiplier of $\mathcal{A}$ such that, given any finite set of seminorms of $\mathcal{A}$, some power of its inverse $e = \Delta^{-N}$ lies in the algebra completion of $\mathcal{A}$ with respect to the seminorms.

Let us give two important examples of algebras satisfying these conditions: firstly the convolution algebra $\mathcal{S}(\mathbf{R})$ with $\Delta = I - \frac{d^2}{dx^2}$, or equivalently (taking the Fourier transform) the multiplication algebra $\mathcal{S}(\hat{\mathbf{R}})$ with $\hat{\Delta}(\xi) = 1+|\xi|^2$; and secondly the algebra of smoothing kernels on the circle, or equivalently the algebra of Schwartz kernels on the real line. (By appropriate choice of orthonormal bases, one sees that both these algebras are isomorphic to the algebra $\mathcal{K}_\infty$ of infinite matrices with coefficients of rapid decay.) There are obvious higher dimensional generalisations of these examples, such as algebras of Schwartz functions on $\mathbf{R}^n$ or smoothing kernels on compact Riemannian manifolds. Of course the latter type of algebra will again be isomorphic to $\mathcal{K}_\infty$, as can be seen by taking an orthonormal basis consisting of eigenfunctions of the Laplacian. To extend Connes' result to multiplier derivations of such algebras, we essentially use the same sort of trick that one uses to make sense of oscillating integrals. There one notices that by formally integrating by parts a certain number of times, an integral that would normally diverge becomes convergent. It is not surprising that similar things work in cyclic cohomology, since this theory may in a certain sense be regarded as highfallutin integration by parts. The trick is also similar to the usual device used for showing that a given kernel operator lies in a certain Schatten class — one introduces an auxiliary operator which is in the same Schatten class but whose inverse when composed with the kernel operator yields a bounded operator.

Proposition 1 $\delta_a^* \phi = b\phi'_a$, **where**

$$\phi'_a(b^1, \ldots, b^n) = \phi(ae, e^{-1}b^1, b^2, \ldots, b^n) - \phi(e, e^{-1}b^1, b^2, \ldots, b^n a).$$

This proposition is easy to verify when a lies in the algebra, and follows in general by continuity (using the approximate identity). It is then an easy matter to deduce as corollaries the strong forms of Morita invariance and additivity that we wished to establish.

II. Use of Outer Derivations: the Poincaré Lemma

All we shall say here has already appeared implicitly in the work of Connes and Goodwillie, but we can give a much quicker and more explicit approach. Again although stated for derivations of an algebra into itself, the results apply more generally — after minor modifications — to derivations of one algebra into another.

Proposition 2 **If ϕ is a cyclic cocycle of the unital algebra $\mathcal{A}$ and δ is a derivation of $\mathcal{A}$, then**

$$S(\delta^*\phi) = b(A\delta_0^t(\phi)),$$

where $\delta_0 : \Omega(\mathcal{A}) \to \Omega(\mathcal{A})$ is the graded derivation of $\Omega(\mathcal{A})$ of degree -1 defined by

$$\delta_0((a^0 + \lambda.1)da^1 \ldots da^{n+1}) = \sum_{j=1}^{n+1}(-1)^{j+1}(a^0 + \lambda.1)da^1 \ldots da^{j-1}\delta(a^j)da^{j+1} \ldots da^{n+1}.$$

This proposition is proved by generalising the formula from differential geometry

$$\mathcal{L}_X = \iota_X d + d\iota_X$$

which expresses the Lie derivative with respect to a vector field X as the graded commutator of the exterior derivative and contraction by the vector field. This identity is at the basis of the proof of the classical Poincaré lemma (see **[13]** and the discussion in Section V). In our case the rôle of the vector field is played by the derivation δ. We have three graded derivations, δ^*, δ_0 and d, on the algebra $\Omega(\mathcal{A})$ of degrees 0, $+1$ and -1 respectively: these play the rôles of Lie derivative (with respect to δ), contraction by δ and exterior differentiation respectively and it is easy to see that their defining formulas are formally the same as the classical analogues. Moreover one checks immediately the validity of the identity

$$\delta^* = d\delta_0 + \delta_0 d. \qquad (\S)$$

Now suppose that ϕ is a cyclic cocycle of $\mathcal{A}$. Then ϕ may be regarded as a closed graded trace $\hat{\phi}$ on $\Omega(\mathcal{A})$ via the formula $\hat{\phi}(a^0 da^1 \ldots da^n) = \phi(a^0, a^1, \ldots, a^n)$. Since $\hat{\phi}$ is closed, we have $d^t\hat{\phi} = 0$, so that from (§) we have $\delta^*\hat{\phi} = d^t(\delta_0^t\hat{\phi})$. The proof of the proposition can then be completed using the following observation, the proof of which follows easily from the definitions.

Lemma **Let $\hat{\psi}$ be a graded trace on $\Omega(\mathcal{A})$, so that $\hat{\phi} = d^t\hat{\psi}$ is a closed graded trace on $\Omega(\mathcal{A})$. Then $S\phi = 2i\pi b(A\psi)$.**

Proposition 2 implies in particular that derivations act trivially on periodised cyclic cohomology. We shall now indicate how to show that they act trivially on the de Rham cohomology $H^n_{DR}(\mathcal{A})$ of $\mathcal{A}$, that is the cohomology of the complex $(H^n(\mathcal{A}, \mathcal{A}^*), I \circ B)$. Recall that the long exact sequence of Connes provides a spectral sequence with E_2 term $H^*_{DR}(\mathcal{A})$, converging to the graded object associated with the periodised cyclic cohomology of $\mathcal{A}$.

Proposition 3 **If ϕ is a Hochschild cocycle of $\mathcal{A}$, define $\delta_1\phi$ by**

$$(\delta_1\phi)(a^0,\ldots,a^{n+1}) = (-1)^n\phi(a^{n+1}a^0,a^1,\ldots,a^n).$$

Then

$$\delta^*\phi = B(\delta_1\phi) + \delta_1(B\phi) + b\psi,$$

where

$$\psi(a^0,\ldots,a^{n-1}) = \sum_{i=0}^{n-1}\sum_{j=0}^{i}(B_0\phi)(a^{i+1},a^{i+2},\ldots,a^{n-1},a^0,\ldots,\delta(a^j),\ldots,a^{i-1}).$$

Thus δ_1 provides a contracting homotopy for δ^* and therefore δ^* induces a trivial action on $H^n_{DR}(\mathcal{A})$. Taking the straightforward generalisations of the above two propositions to derivations from one algebra into another and passing to the integrated versions, one sees that the periodised cyclic cohomology and the de Rham cohomology of a contractible algebra are both trivial. In fact one has more generally that if f_t ($t \in [0,1]$) is a C^1 family of homomorphisms from an algebra $\mathcal{A}$ into an algebra $\mathcal{B}$, then the maps f_t^* all coincide on cyclic cohomology and de Rham cohomology. This generalises a remark of Connes who showed this for cyclic cohomology.

III. Localisation in Hochschild Cohomology: Use of Sheaf Theory

We start by recalling that ϕ is a Hochschild cocycle if and only if $\hat{\phi}(ab\omega) = \hat{\phi}(b\omega a)$ for all $a,b \in \mathcal{A}$ and $\omega \in \Omega(\mathcal{A})$, so that $\hat{\phi}$ satisfies a very weak tracial condition. Now let $\mathcal{C}$ be a central subalgebra of $\mathcal{A}$. For $z \in \mathcal{C}$ we define

$$(\pi_k(z)\cdot\phi)(a^0,\ldots,a^n) = \hat{\phi}(a^0(da^1\ldots da^{k-1})z(da^k\ldots da^n)),$$

where ϕ need now only be a *cochain* and k runs from 1 to $n+1$. The following is the main localisation theorem.

Proposition 4
(1) $b\pi_k(z) = \pi_k(z)b$ for $1 \le k \le n+1$
(2) $\pi_k(z)\cdot\phi$ and $\pi_1(z)\cdot\phi$ are cohomologous for $1 \le k \le n$ if ϕ is a cocycle
(3) $\pi_k(z)\cdot\phi$ is a cocycle whenever ϕ is
(4) $\pi_j(z)$ and $\pi_{j'}(z')$ commute and

$$(\pi_{j_1}(z_1)\cdots\pi_{j_m}(z_m)\phi)(a^0,\ldots,a^n) = \hat{\phi}(a^0(da^1\ldots da^{j_1-1})z_1(da^{j_1}\ldots da^{j_2-1})z_2(\cdots)z_m(da^{j_m}\ldots da^n)),$$

where $j_1 \le j_2 \le \ldots \le j_m$.

Thus the actions π_k all give rise to the same action of C on $H^n(\mathcal{A},\mathcal{A}^*)$. To emphasise the clarity obtained by this explicit approach, we note that if ϕ is a Hochschild cocycle, then the formula

$$\tilde{\phi}(a^0,\ldots,a^n)=\hat{\phi}(a^0da^1.1.da^2.1.\cdots 1.da^n)=(\pi_1(1)\pi_2(1)\ldots\pi_n(1)\phi)(a^0,\ldots,a^n)$$

defines a normalised Hochschild cocycle equivalent to ϕ — we simply observe that $1.d1.1 = 0$ and that the right hand side is equivalent to $\pi_1(1)^n\phi=\phi$, by the proposition.

Let us now illustrate how Proposition 4 may be used to obtain sheaf–theoretic properties of the *duals* of the Hochschild groups $H^n(\mathcal{A},\mathcal{A}^*)$ over the spectrum of C. In the case of Hochschild homology of $\mathcal{A}$ with coefficients in $\mathcal{A}$, these properties are well–known and easy to prove. But if one takes Ext instead of Tor, no general results are known except for the case of noetherian rings. We are working, however, with C^∞ algebras, so it turns out that the appropriate method of localisation is the use of bump functions and partitions of unity within the framework provided by Proposition 4. If we are to work with the duals of Hochschild groups, it is clear from the start that one of the major tricky points will be to show that in the cases under consideration the Hochschild groups are *Hausdorff*, or equivalently that the coboundaries form a closed subspace. Let us now concentrate on one important example to show how one goes about proving such things. Let G be a finite (or more generally compact) group acting on a smooth manifold X and let V be a representation of G. For $\tilde{U}$ open and G–invariant in X, or equivalently for $U=\tilde{U}/G$ open in X/G, we define the *local algebra over* U by $\mathcal{A}(U)=(C^\infty(\tilde{U})\otimes End(V))^G$. We shall simply write $\mathcal{A}$ for $\mathcal{A}(X/G)$ and $H^n(U)$ for $H^n(\mathcal{A}(U),\mathcal{A}(U)^*)$. The two main localisation results are as follows.

Proposition 5 **$H^n(X/G)$ is Hausdorff provided this is true locally, that is if each point in X/G has an open neighbourhood U such that $H^n(U)$ is Hausdorff.**

This follows from Proposition 4. We shall return to the local verification in the next section in the special case when G is finite and V is the trivial representation: we shall need to resort to some fairly involved results from singularity theory. The case when G is finite and $V=\ell^2(G)$ is just the crossed product of $C^\infty(X)$ by G and will be dealt with by a completely different technique in Section VI.

Theorem **Suppose that $H^n(U)$ is Hausdorff for all opens U in X/G. Then $H^n(U)'$ is a reflexive Fréchet space and $U\mapsto H^*(U)'$ defines a complex of fine complete presheaves on X/G with differential provided by $I\circ B$.**

Let us briefly indicate the general techniques used to prove the above results by proving one of the easier parts of the theorem. We will establish the surjectivity of the map

$$\oplus H^n(U_\alpha)\to H^n(U),$$

where $(U_\alpha)_{\alpha\in I}$ is an open cover of the open subset U of X/G. In fact let (χ_α) be a partition of unity subordinate to (U_α) (on $\tilde{U}/G$) and let $\phi \in Z^n(U)$. The continuity of ϕ may be formulated by means of an inequality of the form

$$|\phi(a^0, \ldots, a^n)| \leq C\|a^0\|_{K,N} \cdots \|a^n\|_{K,N},$$

where $K \subseteq U$ is compact and $\|\cdot\|_{K,N}$ is the seminorm given by the supremum over K of the derivatives of order $\leq N$ of a function. Let $\chi \in C_c^\infty(U)$ be a function equal to 1 on a neighbourhood of K, so that $\pi_1(\chi)\cdot\phi = \phi$. Furthermore $\chi = \sum_{\alpha\in F} \chi\chi_\alpha$ for some finite subset $F \subseteq I$ of indices by the compactness of K. Hence we have

$$\phi = \sum_{\alpha\in F} \pi_1(\chi\chi_\alpha)\cdot\phi.$$

Now we take $\chi'_\alpha \in C_c^\infty(U_\alpha)$ such that $\chi'_\alpha(\chi\chi_\alpha) = \chi\chi_\alpha$. Then ϕ is equivalent to

$$\sum_{\alpha\in F} \pi_1(\chi\chi_\alpha)\pi_2(\chi'_\alpha)\cdots\pi_n(\chi'_\alpha)\phi.$$

Since $\chi'(da)\chi' = \chi' d(a\chi'')\chi'$ whenever $\chi''\chi' = \chi'$, it easily follows that each term in the above finite sum for ϕ may be regarded as being defined on the appropriate U_α and thus surjectivity follows.

This theorem can be applied in conjunction with the Poincaré lemma of the previous section to make the local computations of the cohomology of the complex in the statement of the theorem. Indeed this amounts to computing the de Rham cohomology groups $H^n_{DR}(\mathcal{A}(U))$; by the equivariant slice theorem, we only have to consider algebras of the form $(C^\infty(\mathbf{R}^N)\otimes \mathit{End}(V))^H$ where H acts linearly on $\mathbf{R}^N$. By the Poincaré lemma, the inclusion $j : \mathit{End}(V)^H \to (C^\infty(\mathbf{R}^N)\otimes \mathit{End}(V))^H$ induces an isomorphism on de Rham cohomology.

Similar results apply for algebras of smooth functions on orbifolds. An orbifold is locally described by the quotient of a euclidean space by the linear action of a finite group. (For detailed definitions and properties of orbifolds we refer the reader to the original work of Satake **[18]** on 'V–manifolds' and Chapter X of **[16]**.) As Haefliger has pointed out, using the fact that the (orthonormal) frame bundle is actually a manifold, one can describe an orbifold as the quotient of a smooth manifold by a compact group with finite isotropy subgroups. The smooth functions are just the invariant functions on the manifold and thus the local algebras are exactly the same as the case of an orbit space. We note that it is not always possible to compute the local Hochschild groups explicitly, indeed in many cases they can grow inordinately large far beyond the dimension of the underlying manifold X: a similar phenomenon has been shown to occur by Feigin and Tsygan **[8]** for coordinate rings of affine varieties that are not complete intersections. (We should also perhaps refer the reader to the parallel work of Masuda and Natsume **[15]** on the cyclic cohomology of zero dimensional affine schemes.) We will explain the link between these two examples in the next section. In propitious circumstances, however, where the Hochschild cohomology is decently behaved and has a good geometric candidate, one can use the theorem to show that it is indeed given by what one expects just by doing a local verification. Such is the case for crossed products or when one has an orbifold where the isotropy groups are finite reflection groups (that is, generated by reflections). These will be discussed in Sections V and VI. (We have not yet considered the case when the orbifold is locally a complete intersection; although by the work of Kac–Watanabe **[12]** and Nakajima **[17]**, one has essentially complete knowledge of the restrictions on the isotropy groups.)

IV. Local Computations: Transition from Algebraic to C^∞ Resolutions using Singularity Theory

We wish to study the algebra $C^\infty(V)^G$, where V is a euclidean space endowed with a linear action of G, and we would like to do this by using the smaller subalgebra $\mathbb{C}[V]^G$ of invariant polynomials. This is permitted by the following generalisation of a result of Malgrange. It enables one to 'boot–up' resolutions of the $\mathbb{C}[V]^G$ (as a bimodule over itself) to obtain resolutions for $C^\infty(V)^G$.

Proposition 6 $C^\infty(V)^G$ **is flat over** $\mathbb{C}[V]^G$**.**

Now by Hilbert's theorem, $\mathbb{C}[V]^G$ is the coordinate ring of an affine variety — this observation and Proposition 6 therefore provide the promised explanation of the link with the work of Feigin and Tsygan. It accordingly has a (pas–a–pas) resolution by free $\mathbb{C}[V \times V]^{G\times G}$–modules of finite rank. Unless G is a finite reflection group, however, this resolution will have infinite length. Indeed one knows, from the work of Quillen and André on the homology of rings, that one expects the resolution to be 'finitely generated' (as an algebra) only if the associated affine variety is locally a complete intersection. At any rate, this resolution yields a resolution of $C^\infty(V)^G$ by free $C^\infty(V \times V)^{G\times G}$–modules of finite rank with the differentials given by matrices of invariant polynomials. The Hausdorff property of the Hochschild cohomology of $C^\infty(V)^G$ is then a direct consequence of the following effective version of one of the basic results in the theory of partial differential equations with constant coefficients (see **[14]**).

Proposition 7 (Existence of Extensors for Polynomial Maps) **Let** $\Omega \subseteq \mathbb{R}^n$ **be open and** $P = (P_{ij})$ **be a matrix with polynomial entries defining a module map** $P : C^\infty(\Omega)^p \to C^\infty(\Omega)^q$**. Then there is a continuous** $\mathbb{C}$**–linear map**

$$T : P(C^\infty(\Omega)^p) \to C^\infty(\Omega)^p$$

such that $TP = id$**. Hence** $im(P)$ **is closed and** $ker(P)$ **is complemented (over** $\mathbb{C}$**).**

We found a relatively short proof of this result using the division theorem of Malgrange and Mather (see [19]) and a special version of Weierstrass' preparation theorem due to Hörmander [11]. One simply has to adapt part of Hörmander's proof of Oka's Theorem with bounds as it appears in his treatment of the L^2 cohomology of the $\overline{\partial}$ equation. Subsequently we discovered that E. Bierstone and G. Schwarz in [2] had already proved a far more general version of Proposition 7 by a quite different method. Although their proof is much longer, it does use an interesting principle from the theory of topological vector spaces which in certain circumstances allows one to show that resolutions that are known to be exact must automatically be allowable in the sense of Mac Lane or admissible in the sense of Connes. (We recall that a topological resolution is allowable provided that all the kernels are complemented.) In our case, the fact that the kernels are complemented follows directly from Proposition 7. The principle of Vogt states that a subspace of a Fréchet space will be complemented provided that the subspace is isomorphic to a quotient of $\mathcal{S}$ and that the quotient by the subspace is isomorphic to a closed subspace of $\mathcal{S}$. Here $\mathcal{S}$ denotes the universal nuclear space of sequences of rapid decay. It follows by considering eigenfunction expansions of the Laplacian, that the smooth functions on a compact manifold without boundary are isomorphic to $\mathcal{S}$; whereas if one allows the manifold to be non–compact or to have non–empty boundary, one just gets a quotient of $\mathcal{S}$. Thus the principle cannot immediately be applied to modules over $C^\infty(\Omega)$, which gives an indication as to why the proof of Bierstone and Schwarz is so involved. Besides, it is more in the spirit of Malgrange's (non–extensor) version of the division theorem rather than Mather's later effective formulation and proof.

As a result of these local computations, one can show that the following 'invariance principle' applies.

Theorem **The periodised cyclic cohomology of $C^\infty(X)^G$ is isomorphic to the G–invariant part of the periodised cyclic cohomology of $C^\infty(X)$: $H^*(C^\infty(X)^G) = H^*(C^\infty(X))^G$.**

Of course, much more can be said. Although the Hochschild cohomology groups cannot effectively be computed, one knows that the de Rham cohomology is already equal to the graded object associated with the peiodised cyclic cohomology and may be identified with the invariant part of the de Rham homology of X, which is nothing but the de Rham homology of X/G. In short the periodised cyclic cohomology of the invariant functions can be identified with the de Rham homology of the quotient space. The same reasoning shows that the periodised cyclic cohomology of the algebra of smooth functions on an orbifold can be identified with the de Rham homology (defined using currents) of the orbifold. To obtain more precise results on the Hochschild cohomology groups and unperiodised cyclic cohomology, one has to place restrictions on the isotropy subgroups. For example one could demand that the affine varieties corresponding to the 'local' algebras $C^\infty(T_x(X))^{G_x}$ are all smooth. We shall adopt this assumption in the next section since it is what we require for the computations in our main example $\mathcal{S}(G)$, although as we have already mentioned it would be equally interesting to relax this condition and just ask for complete intersections locally.

V. Invariant Theory for Finite Reflection Groups

When G is a finite reflection group acting on the finite dimensional real vector space V, there are three important results concerning the ring of invariant polynomials $\mathbb{C}[V]^G$ (see [4]).

1. *Newton's Theorem:* $\mathbb{C}[V]^G = \mathbb{C}[p_1, \ldots, p_n]$ where n is the dimension of V. Thus the invariant polynomials themselves form a polynomial ring in n fundamental invariants $p_1, \ldots, p_n$. The fundamental invariants can be given explicitly by extending ideas of R. Steinberg.
2. *Chevalley's Theorem:* $\mathbb{C}[V]$ is free of rank $|G|$ as a module over $\mathbb{C}[V]^G$.
3. *Solomon's Theorem:* $(\mathbb{C}[V] \otimes \Lambda(dV))^G = \mathbb{C}[p_1, \ldots, p_n] \otimes \Lambda(dp_1, \ldots, dp_n)$. This statement is equivalent to the statement that the G–invariant (algebraic) k–forms on V all arise as linear combinations of forms of the form $f^0 df^1 \wedge \cdots \wedge df^k$, where the f_j's are invariant polynomials on V.

All these results extend in a natural way to the C^∞ setting, as a consequence of the work of Whitney, Malgrange, Glaeser, Mather and others. To see the inherent difficulties here, one should try to work out how to define $\partial f / \partial p_k$ when f is an invariant smooth function on V. After one has the C^∞ results at one's disposal, one can then freely use the Koszul resolution (described below) to perform the explicit computation of the Hochschild cohomology of $C^\infty(V)^G$ and hence to compute the Hochschild and cyclic cohomology of an orbifold for which the isotropy groups act as finite reflection groups on the tangent space. We will call such an orbifold a *Chevalley* orbifold. Let us now summarise the results on the various cohomology groups of the algebra of smooth functions on a Chevalley orbifold. These provide the promised refinements of the results in the previous section.

Theorem **Let X be a Chevalley orbifold. The Hochschild cohomology groups of $C^\infty(X)$ are given by the spaces $\mathcal{D}_n$ of currents on X with the operation $I \circ B$ corresponding to the de Rham boundary operator d^t. Thus the de Rham cohomology of $C^\infty(X)$ is given by the de Rham homology of X. The cyclic cohomology of $C^\infty(X)$ is given by**

$$H_\lambda^n(C^\infty(X)) = ker(d^t) \cap \mathcal{D}_n \oplus H_{n-2}(X, \mathbb{C}) \oplus H_{n-4}(X, \mathbb{C}) \oplus \cdots.$$

We also have a refinement of the invariance principle of Section IV.

Theorem **Let X be a smooth manifold and let G be a finite group acting smoothly on X, such that the stabilisers G_x all act as reflection groups on T_xX. Then the Hochschild cohomology, the de Rham cohomology, the cyclic cohomology and the periodised cyclic cohomology of $C^\infty(X)^G$ can all be canonically identified with the G–invariant parts of the corresponding groups for $C^\infty(X)$.**

We shall now give three applications of the above theorems. Firstly if we take a trivial action, we obtain yet another method of determining the Hochschild cohomology of $C^\infty(X)$ when X is a smooth manifold (not necessarily compact) without exhibiting an explicit global resolution of $C^\infty(X)$ as a $C^\infty(X\times X)$–module. This provides an alternative, perhaps simpler, method to that of Connes, although the main tool — Connes' formula for the contracting homotopy in the Koszul resolution — is common to both approaches.

For the benefit of the reader we briefly recall some details of the Koszul complex along with the de Rham complex, which may be regarded as the 'dual' complex. Let $\mathcal{A} = \mathbb{C}[V]$ and let $\Omega^*_{\mathcal{A}}$ be the algebra of forms on $\mathcal{A}$. We are interested in finding a resolution for the diagonal map $\mathcal{A}\otimes\mathcal{A}\to\mathcal{A}$ by projective or even free $\mathcal{A}\otimes\mathcal{A}$–modules. Using the obvious reparametrisation of $V\times V$ given by taking the diagonal and a transverse copy of V, this is equivalent to resolving the augmentation map $\varepsilon:\mathcal{A}\to\mathbb{C}$, $\quad f\mapsto f(0)$. To define the complex, we introduce the $GL(n)$–invariant vector field $X = x_1\frac{\partial}{\partial x_1}+\cdots+x_n\frac{\partial}{\partial x_n}$ on V together with the flow $\alpha_t(x)=tx$. The Koszul and de Rham complexes are respectively

$$0\to\Omega^n_{\mathcal{A}}\xrightarrow{\iota_X}\Omega^{n-1}_{\mathcal{A}}\xrightarrow{\iota_X}\cdots\xrightarrow{\iota_X}\Omega^1_{\mathcal{A}}\xrightarrow{\iota_X}\Omega^0_{\mathcal{A}}=\mathcal{A}\xrightarrow{\varepsilon}\mathbb{C}$$

and

$$\mathbb{C}\to\mathcal{A}=\Omega^0_{\mathcal{A}}\xrightarrow{d}\Omega^1_{\mathcal{A}}\xrightarrow{d}\cdots\xrightarrow{d}\Omega^{n-1}_{\mathcal{A}}\xrightarrow{d}\Omega^n_{\mathcal{A}}\to 0.$$

These complexes are exact with contracting homotopies S and k given by $S\omega = \int_0^1\alpha_t^*(d\omega)/t\,dt$ and $k\omega=\int_0^1\alpha_t^*(\iota_X\omega)/t\,dt$ respectively. Thus $S\iota_X+\iota_XS=1$ and $kd+dk=1$; these formulas follow from the identity $\iota_Xd+d\iota_X=\mathcal{L}_X$ and the formula $\frac{d}{dt}\alpha_t^*\omega = \frac{1}{t}\alpha_t^*\mathcal{L}_X\omega$.

Our approach replaces Connes' somewhat ad hoc method of globalising the Koszul resolution by the systematic use of the sheaf theoretic properties of the (duals of the) Hochschild cohomology groups. The advantage of the latter approach is that it applies to far more general situations where it is not clear that one can find global resolutions. Our second application is a case in point. Here one can use the local computations for certain rather simple Chevalley orbifolds to compute the Hochschild cohomology of the algebras of smooth functions on manifolds with boundaries. (This method is equally applicable to manifolds with more exotic boundaries, for example where the boundary is piecewise linear (i.e. has corners) or is polyhedral.) The idea here is that in a neighbourhood of any point on the boundary, the smooth functions correspond to the smooth functions on a half–space $\mathbb{R}^{n-1}\times\mathbb{R}_+$. Using the map $(x_1,\ldots,x_{n-1},x_n)\mapsto(x_1,\ldots,x_{n-1},x_n^2)$, this algebra may be identified with the smooth functions on $\mathbf{R}^n$ invariant under the reflection $x_n \mapsto -x_n$. Since we know how to work with this latter algebra, we also know how to work with the algebra of smooth functions on a half–space. Our third and final example was the one we needed for the Schwartz space calculations. Let G be a reflection group on V and make G act on $W = V \oplus \mathbf{R}$ by taking the trivial action of G on $\mathbf{R}$. Let S be the unit sphere in W; then the action of G on S is locally by reflection groups and therefore the invariance principle is applicable.

VI. Crossed Products by Finite Groups

Let G be a finite group acting by automorphisms on an algebra $\mathcal{A}$ via an action $\alpha : G \to Aut(\mathcal{A})$. The crossed product $\mathcal{A} \rtimes G$ may either be defined as the algebra of formal sums $\sum_{g\in G} a_g g$ subject to the rule $gag^{-1} = \alpha_g(a)$; or equivalently as the fixed point algebra $(\mathcal{A} \otimes End(\ell^2(G)))^{\alpha\otimes Ad(\lambda)}$, where λ denotes the left regular representation of G on $\ell^2(G)$. Now it is fairly easy to see that if $(\mathcal{M}_k, d)$ is a projective resolution of $\mathcal{A}$ as an $\mathcal{A}$–bimodule which is equivariant for the *diagonal* action of G and $i_*\mathcal{M}_k = ind_{G\uparrow G\times G}\mathcal{M}_k$, then $(i_*\mathcal{M}_k, i_*d)$ will provide a projective resolution for $\mathcal{A} \rtimes G$ as a bimodule over itself. Taking the standard cobar resolution $(A^{\otimes^n})$, we obtain the following formula for the Hochschild cohomology of the crossed product:

Proposition 8 $H^n(\mathcal{A} \rtimes G, (\mathcal{A} \rtimes G)^*) = \oplus_{g\in conj(G)} H^n(\mathcal{A}, \mathcal{A}_g^*)^{C(g)}$.

Here the sum is over the conjugacy classes of G, $C(g)$ denotes the centraliser of g and $\mathcal{A}_g^*$ is the bimodule A^* with the action of $\mathcal{A}$ on the right perturbed by g. Note that Proposition 8 allows one to compute the Hochschild cohomology groups of the crossed product using a projective resolution of $\mathcal{A}$ which has no compatibility with G. In particular if one wishes to compute the cyclic cohomology of $C^\infty(X) \rtimes G$, where X is a smooth manifold on which G acts by diffeomorphisms, one can use any old projective resolution of $C^\infty(X)$ as a module over $C^\infty(X \times X)$ in Proposition 8. We leave it as an exercise for the reader to work out what happens if one uses Connes' global resolution to do this computation; his or her answer should hopefully tally with the result below. Let us, however, show how to proceed using the localisation methods of Section III. Again we find that the duals of the Hochschild groups form presheaves over the orbit space X/G. So to determine the Hochschild cohomology globally, we only have to make an educated guess as to what it might be and then do a local check on algebras of the form $C^\infty(V) \rtimes G$. Here we may use the Koszul resolution (which incidentally is equivariant for the diagonal action of G). Thus it is much easier to perform the computations for the crossed product than for the fixed point algebra where local resolutions were quite hard to produce. We summarise our findings.

Theorem **Let G be a finite group acting by diffeomorphisms on a smooth manifold X. The Hochschild cohomology groups $H^n(\mathcal{A},\mathcal{A}^*)$ of the crossed product $\mathcal{A} = C^\infty(X) \rtimes G$ may be identified with the direct sum of the spaces of invariant currents on the fixed point submanifolds $X^g = \{x \in X : gx = x\}$**

$$H^n(\mathcal{A},\mathcal{A}^*) = \oplus_{g \in conj(G)} \mathcal{D}_n(X^g)^{C(g)},$$

with the operation $I \circ B$ corresponding to the de Rham boundary operators on the currents on the appropriate fixed point manifolds. The de Rham cohomology of $\mathcal{A}$ thus is given by the sum of the de Rham homology groups of the orbifolds $X^g/C(g)$ as g runs over the conjugacy classes of G and coincides with the graded object associated with the periodised cyclic cohomology of $\mathcal{A}$. The cyclic cohomology of $\mathcal{A}$ is given by

$$H^n_\lambda(\mathcal{A}) = \oplus_{g \in conj(G)} H^n_\lambda(C^\infty(X^g))^{C(g)}.$$

It is of course interesting to do the analogous computations for the crossed product by a compact group, even just a torus. For cyclic homology related computations have been made by J. Brylinski [5] and J. Block [3]. Their methods, however, do not extend immediately to yield analogous results for cyclic cohomology, which is less well behaved. For example Block considers not the full topological crossed product $\mathcal{A} \rtimes G = (\mathcal{A} \hat{\otimes} \mathcal{K}_\infty(G))^{\alpha \otimes Ad(\lambda)}$ but instead the subalgebra corresponding to the algebraic crossed product. Thus $\mathcal{K}_\infty(G)$ is replaced by the algebra of finite rank operators on $L^2(G)$ corresponding to finite dimensional G–invariant subspaces V of $L^2(G)$ and the algebraic crossed product therefore becomes the inductive limit of the algebras $(\mathcal{A} \otimes End(V))^G$. This is perfect for cyclic homology since — like K–theory — it commutes with inductive limits. The same is not true for either K–homology or cyclic cohomology for which, on the contrary, one has good behaviour for projective limits of topological algebras, as we shall see in the final section of this paper. Thus there is still work to be done. It should be pointed out, however, that already in 1984 Connes had made extensive computations for the cyclic cohomology of crossed products by the circle group.

VII. Use of Continuity: Schwartz Algebras on $\mathbb{R}^N$

We have already shown in Section I how continuity conditions on cocycles can be used to prove that multiplier derivations act trivially. We shall now give another example which again uses the idea of extending a cocycle by continuity to a larger algebra where it can be treated with greater ease. We start by recalling the classical fact that, by means of stereographic projection, $\mathcal{S}(\mathbb{R}^N)$ can be identified with the subalgebra of $C^\infty(S^N)$ consisting of all those functions which vanish to infinite order at some given point (in fact, the point of projection). It is clear that all problems revolve around what happens at this point where infinitely many conditions are imposed on a function and its derivatives. Now any cyclic cocycle will have its continuity specified by a seminorm involving only finitely many derivatives and hence will extend by continuity to the algebra of smooth functions on S^N whose first k derivatives (say) vanish at the given point. It is convenient, however, to make a more judicious choice of extending algebra, since the obvious largest smooth algebra to which the cocycle can be extended is not as tractable as one would like. Indeed if one were to work with it, one would get involved with problems of nilpotent algebras as soon as one tried to determine the parameters which measured the obstructions to extending the cocycle to the whole of $C^\infty(S^N)$. Thus although this algebra provides a possible line of attack, there is a nicer intermediate algebra with which to work, namely the algebra of functions on S^n whose Taylor series at the point has only homogeneous terms of total degree a multiple of some fixed odd integer $2k+1$. (Here we take an obvious chart for the hemisphere centred on the point of projection, making the point go to the origin.) The intermediate algebra can then be understood by means of the following elementary observation.

Proposition 9 **Let k be a positive integer. Then the map $\gamma_k : x \mapsto \|x\|^{2k} \cdot x$ induces a topological isomorphism of $C^\infty(\mathbf{R}^N)$ onto the subalgebra consisting of functions whose Taylor series at 0 have only homogeneous terms of total degree a multiple of $2k+1$.**

Now using a bump function, one can construct an increasing function $\psi \in C^\infty([0,\infty))$ such that $\psi(t) = t$ for small t and $\psi(t) = 1$ for large t. One can then use the modified map $\gamma_k' : x \mapsto \psi(\|x\|^{2k}) \cdot x$ to construct an isomorphism analogous to that of Proposition 9 for functions on S^N rather than on $\mathbf{R}^N$. In this way the cyclic cohomology of $\mathcal{S}(\mathbf{R}^N)$ can be deduced from known results on the cyclic cohomology and Hochschild cohomology of $C^\infty(S^N)$. In performing this computation, it should be born in mind that the algebra of smooth functions on $\mathbf{R}^N$ vanishing to infinite order at 0 is invariant under multiplication by arbitrary powers of $\|x\|$. There is a corresponding result for the analogous algebra on S^N. Furthermore the maps γ_k and γ_k' induce automorphisms of these algebras.

Theorem **For $n \leq N$, $H^n_\lambda(\mathcal{S}(\mathbf{R}^N)$ coincides with the space of closed tempered currents on $\mathbf{R}^N$. For $n = N + 2k$ with $k > 0$, the group is one–dimensional, generated by $S^k\phi$ where ϕ is the fundamental N–trace**

$$\phi(f^0, f^1, \ldots, f^N) = \int_{\mathbf{R}^N} f^0 df^1 \wedge \cdots \wedge df^N.$$

For $n = N + 2k + 1$ with $k \geq 0$, the cyclic cohomology groups vanish.

References

1. J. Arthur, *A theorem on the Schwartz space of a reductive Lie group*, Proc. Nat. Acad. Sci. U.S.A., **72**, no. 12 (1975), 4728–4719.
2. E. Bierstone and G. W. Schwarz, *Continuous linear division and extension of C^∞ functions,* Duke Math. J. **50** (1983), 233–271.
3. J. Block, Thesis, Harvard University (1987).
4. N. Bourbaki, "Groupes et Algèbres de Lie," IV–VI, Hermann, 1968.
5. J.–L. Brylinski, *Algebras associated with group actions and their homology*, preprint, Brown University (1987).
6. A. Connes, *de Rham cohomology and non–commutative algebra*, Publ. Math. I.H.E.S. **62** (1986), 94–144.
7. A. Connes and H. Moscovici, *The L^2–index theorem for homogeneous spaces of Lie groups*, Ann. of Math. **115** (1982), 291–330.
8. B. Feigin and L. Tsygan, *Additive K–theory and crystalline cohomology*, J. Funct. Anal. Appl. **19** (1986), 124–132 (english translation).
9. E. Getzler, *Pseudodifferential operators on supermanifolds and the Atiyah–Singer index theorem,* Comm. Math. Phys. **92** (1983), 163–178.
10. T. G. Goodwillie, *Cyclic homology, derivations, and the free loop space*, Topology **24** (1985), 187–215.
11. L. Hörmander , "An introduction to complex analysis in severable variables," North Holland, 1973 and 1979.
12. V. Kac and K. Watanabe, *Finite linear groups whose ring of invariants is a complete intersection*, Bull. Amer. Math. Soc **6** (1982), 1221–1223.
13. S. Lang, "Differentiable Manifolds," Addison–Wesley, 1972.
14. B. Malgrange, "Ideals of Differentiable Functions," O.U.P., Bombay, 1966.
15. T. Masuda and T. Natsume, *Cyclic cohomology of certain affine schemes*, Publ. R.I.M.S. **21** (1985), 1261–1279.
16. J. W. Morgan and H. Bass (eds.), "The Smith Conjecture," Academic Press, 1984.
17. H. Nakajima, *Quotient singularities which are complete intersections*, Manuscripta Math. **48** (1984), 163–187.
18. I. Satake, *On a generalization of the notion of manifold*, Proc. Nat. Acad. Sci. U.S.A. **42** (1956), 359 -363.
19. C. T. C. Wall (ed.), "Proceedings of Liverpool Singularities Symposium — I," Lect. Notes in Math, no. 192, Springer–Verlag, 1971.
20. A. J. Wassermann, *Une démonstration de la conjecture de Connes–Kasparov pour les groupes de Lie linéaires connexes réductifs*, C.R.A.S. **304** (1987), Série I, no. 18, 559–562.
21. A. J. Wassermann, *Cyclic Cohomology I: Finite Group Actions and Schwartz Algebras*, preprint, University of Liverpool (1987).
22. A. J. Wassermann, *Cyclic Cohomology II: A Generalisation of the Connes–Moscovici Index Theorem*, in preparation.